The Transfer of Knowledge Through Art and Visualization

This book calls attention to the significance of visual communication of knowledge, collaborative working style, and our openness to novelty needed in our schools, workplaces, industry, and elsewhere. The book's leading theme is an analysis of strategies conducive to inclusive learning and instruction: sketching, visualizing, visual storytelling, animating, and serious gaming. Visual solutions can grasp the learners' and users' attention and novelty comprehension.

This book addresses people who seek new strategies in studying and working environments by learning, presenting, and marketing their work visually. The readers this book addresses are teachers, employers, recruiters, and instructors in institutions and companies, business coaches, marketers, and managers, along with employees belonging to recent generations focused on starting their future companies and students learning at schools and colleges who build their portfolios to prepare for job applications and further careers.

The readers can benefit when solving learning projects from tools enhancing their visual and technological literacy and inventive style in learning and working. The book also offers short activities supporting the text. The readers can practice presenting knowledge visually, collaborating with science specialists, and learning novel technologies as tools for delivering knowledge transfer. With the novelty of advanced technologies and artificial intelligence–based techniques, we all need these skills.

Anna Ursyn is Professor at the University of Northern Colorado, USA. Anna has had over 50 single juried and invitational art shows, participated in over 200 fine art exhibitions including musea, such as over a dozen times at the ACM SIGGRAPH Art Galleries, and travelling shows, Louvre, Paris, NTT Museum in Tokyo (5,000 texts and 2,000 images representing XX Century), and Virtual Media Network, (the largest moving-image outdoor display, Dallas Texas). Since 1987, she also serves as a liaison, Organizing and Program Committee member, of International IEEE Conferences on Information Visualisation (iV) London, UK. She also serves as a chair of the Symposium and Digital Art Gallery D-ART iV. Her artwork was selected to be sent to the moon by NASA as a part of the MoonArc Project by Carnegie Melon University and travelling shows including Centre Pompidou, Paris. Her work in the ABAD exhibition is in the permanent collections of the Museum of Modern Art in New York, the Los Angeles County Museum of Art, permanent collection of Museé de la Poste in Paris, France, and the Smithsonian Institution.

The Transfer of Knowledge through Art and Visualization

Novel Technologies, Collaboration, and Visual Communication of Science-based Projects

Anna Ursyn

Routledge
Taylor & Francis Group

NEW YORK AND LONDON

First published 2024
by Routledge
605 Third Avenue, New York, NY 10158

and by Routledge
4 Park Square, Milton Park, Abingdon, Oxon, OX14 4RN

Routledge is an imprint of the Taylor & Francis Group, an informa business

ISBN: 978-1-032-62422-8 (hbk)
ISBN: 978-1-032-70533-0 (pbk)
ISBN: 978-1-032-70534-7 (ebk)

DOI: 10.4324/9781032705347

Typeset in Times New Roman
by Apex CoVantage, LLC

Content

Figures

Tables

1 Introductory premises and assertions

Introduction

It is hard to imagine a workspace that does not involve a collaborative spirit. We deliver together. Cooperation and teamwork, an integrative approach to learning and working, visual and technological literacy, and openness to novelty are of growing significance in our environments, such as schools, workplaces, industry, and elsewhere.

The growing share of artificial intelligence (AI) – based services in our life provides us with a tool for producing visual materials. However, the quality of our writings, artworks, and professional products depends on our choice of prompts for the artificial intelligence–based apps. Depending on the prompts, the AI output may enhance and enrich our projects or repeat common trends and fashionable styles.

Visual solutions can grasp the readers' and users' attention and novelty comprehension. Three figures presented here illustrate the changing spectrum of possible visual presentations, from simple computer graphics on first personal computers (Figure 1.1) through coding algorithmic programs (Figure 1.2) to pictures supported by AI and scientific material of a nanostructure (Figure 1.3). We can see the corresponding changes in presenting objects through visualizations, advertising, and marketing.

Learning projects are collaborative or individual, technology based, and science inspired. Visual-verbal science-based projects may take the form of interactive collaboration on narrative visualizations, including visualization techniques in science learning and instruction. Projects offer challenging tasks because they are engaging both intellectually and visually. Visual storytelling and visualization exercises engage participants in the interdisciplinary field of the visual way of presenting science-related themes. Part of the project aims to hone the readers' artificial intelligence literacy, visual literacy, and computer graphics skills. Art-based projects support knowledge delivery and provide motivation and inspiration, expanding one's field of interest and prospective specialty. Projects encourage learners to look for supporting

DOI: 10.4324/9781032705347-1

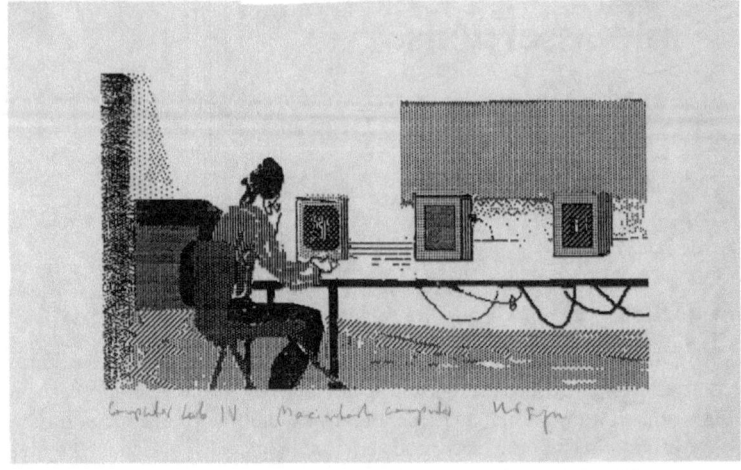

Figure 1.1 An old computer lab

Figure 1.2 Report from Colorado

information from other areas of knowledge. They are built for instructors and learners in schools and academies and members of staff in business and marketing companies, which train their workers to present content (entertainment, promotion, social media, marketing, and more) through the arts and visual media.

Figure 1.3 A Nano Colony (illustration supported by AI and scientific material of nano-structures provided by Cris Orfescu)

The aim of this book

This book addresses people who seek new visual ways of learning, presenting, and marketing their work. It offers valuable tools for current employees, prepares students for future careers, and supports learning and instruction in a visual way. The book aims to develop the readers' interest in tasks vital in almost all kinds of work: learning about new technologies requiring conceptual thinking, working in a collaborative style, and evolving visual communication of knowledge. This book supports the transition of readers from passive recipients of knowledge to active creators of novel ideas, concepts, and knowledge.

To whom this book may concern

Developments in the media and artificial intelligence technologies made visual ways of communication crucial. The ability to communicate content effectively is necessary not only for knowledge transfer but for employees' training, writing user manuals, marketing, recreation, entertainment, and many other activities. The book is addressed to many groups of readers who can profit from the visual communication of content.

1   Teachers can find descriptions of visual projects designed for present-day generations of audiences.
2   Students will find preparation for applying successfully with a visually strong portfolio for future work as creative, opened to art professionals and technicians working in new areas of knowledge.
3   Recruiters in institutions and companies need to know how to search for candidates and applicants with solid knowledge of visual media and visual communication skills.
4   Trainers, business coaches, and employers in institutions may need to update their skills in visual media to communicate with their employees and hone their skills in reaching their clients and users.
5   Marketers and managers, who need to address the current audience successfully, may seek new visual ways to promote and advertise their work. To connect well with their clients, they need to know current buzzwords in media and technologies.
6   Readers belonging to the Millennial generation who own their own companies and young Generation Z members focused on starting future companies make now a significant part of the audience. While focused on their careers, they will later hone these valuable skills.

The importance of the collaborative style and visual communication of knowledge

Currently, specialists that work in various fields of knowledge and different countries are developing new technological solutions. This fact creates a need for a collaborative spirit, inclusiveness, sharing novelty, and cooperation with international scope. Learning about coding and computer programs is needed because it improves communication with technology-savvy specialists. Moreover, knowledge about coding helps to understand how artificial intelligence bots work and interact with autonomous systems or programs on the internet and networks.

This book focuses on cooperation and novelty understanding because they are becoming essential needs that must be satisfied. Elisabeth Kolbert (2023) listed the world's urgent problems: the world's phosphorus problem, "the carbon-dioxide problem, its plastic problem, its soil-erosion problem, its groundwater-use problem, its soil-erosion problem, and its nitrogen problem." Each of these vital problems requires both devising novelty solutions and the collaboration of specialists in many disciplines.

AI, its benefits and perils in our technological landscape

Electronic, non–protein-based intelligence (or memory) develops faster than expected. Thus collaborations with machine intelligence are constantly expanding. In response to a human prop, AI machines draw from a variety of disciplines, subdisciplines, and subsubdisciplines that are present in many fields and on many levels of knowledge, in terms of philosophy, social science, semiotics, and language arts along with sciences, computing, technologies, and other sources that are unknown for us. Moreover, an AI machine remembers everything it learned till now and improves itself when it reiterates versions of its answers even without changing the initial prompt by the client. This way, the AI output's progress and value are immense and unforeseeable.

Despite all the ways artificial intelligence can benefit the world, AI experts fully recognize the possible dire implications of AI developments and the significance of risks they may bring about. Everybody may have an AI machine to cause irreparable harm to humanity. For this reason, despite the uncontrolled growth of AI's impact on our reality, people are working on defining the limits within which it can be used, shaping new legislation, and formulating new laws. To reduce the societal-scale risks from AI, a group of about one hundred AI experts founded the Center for AI Safety (CAIS, pronounced 'case'), a San Francisco-based research and field-building nonprofit. On June 2, 2023, the AI experts posted on the CAIS website (*www.safe.ai*) a one-sentence statement on AI risk: "Mitigating the risk of extinction from AI should be a global priority alongside other societal-scale risks such as pandemics and nuclear war." The urgency of undertaking work on securing the safety of humans has been recognized in other regions of the world.

This book examines a new form of cooperation, this time between humans, machines, and again humans, which is occurring now on several levels. Human participation occurs at the initial and final phases of a task or project. (1) A user, as a prompt writer, monitors the machine's work; (2) an AI machine performs the task and offers its solutions as an output; (3) the user chooses and optimizes the result, still with the help of AI. Other human team members often revise the outcome. Learning about cooperation with generative AI-based software provides users with benefits. For example, one can improve one's writing technique: first, the OpenAI ChatGPT generates a paper according to the author's prompt, then the author edits the text by adding prompts for ChatGPT, adds comments and perspectives, and improves the writing quality and the choice of words. Also, one can facilitate knowledge transfer and explain complicated concepts by creating videos using ChatGPT.

A need for technological literacy

Visual ways of learning, presenting, and marketing create a demand for schooling in new areas of knowledge. There is high demand for skilled professionals and technicians in the industry. Business coaching is a fast-growing discipline helping companies work more intelligently and likable. Programmers can co-operate with users when they distribute software for many computing devices, thus securing collaboration. For example, they compute machine control modeling, drone data analysis, coordinate efforts of engineers and builders, control by satellite field watering and lawn sprinkler systems depending on the season, and create satellite-based navigation of self-driving cars. Education with art should support the development of abstract thinking. Abstract thinking abilities are essential in both life and education: in mathematics, philosophy, poetry, art, and science.

Advancements in networking are resulting in the expansion of visual communication. They should be included in instructional strategies and instruction about new solutions in computer graphics.

Many people feel it is important to expand the role of computer graphics and coding from early childhood, not waiting till learners become young adults or adults. Perhaps, teaching very young children to play chess and understand Latin or some Asian language may provide different cognitive challenges that could prepare them to grasp complex, difficult, and, to some people, scary concepts behind computing and understand various tactics and sets of routines.

A need for visual literacy

The recurring mantra in this book is the growing importance of visuals and visual presentation of knowledge. Presenting numbers and numeric data in a visual way adds meaning. It changes the cryptic, hard-to-understand data into revealing information and thus makes our visual insight and perception stronger. Visualization and visual metaphors (that symbolically represent other, often abstract concepts) are omnipresent in industry, business, commerce, marketing, and advertising. Sciences overlap, as the arts do.

For example, it is hard to describe an atom without discussing physics, chemistry, math, geometry, or art. At the same time, the growing access to technology makes it easier for an artist to learn and collaborate with science-oriented people. The same can be said about a coder. This book advises developing habits supporting visual ways of learning, including making visual shortcuts for concepts and processes under study.

Many students learn easier when they draw and sketch what they are learning. To study physics, one may prefer to draw graphs when learning equations related to concepts such as velocity or acceleration or drawing diagrams (with graphical symbols for the electric current, resistance, capacity, or impedance) to understand the relations between laws of electricity better. Learning chemical reactions may be more accessible while one draws 2D or 3D models of compounds and their chemical reactions using symbols for elements and their valences (for example, a skeletal, ball-and-stick, or a space-filling model for a planar ring of a chemical compound, benzene, along with its chemical formula

C_6H_6). The same way of learning biochemistry may ease understanding reactions between sugars or the 22 alpha amino acids appearing in the genetic code of living beings (using symbols for hydrogen, oxygen, nitrogen, and their side chains). While studying anatomy with great numbers of English and Latin names, it may be easier to learn them by sketching individual bones or tissues and their configurations in the whole organism. For geography and history, many students get help while drawing maps of physical, political, and historical events while they memorize names and dates. Drawing graphical visualizations such as plots, bar or pie charts, diagrams (e.g., Venn diagrams with overlapping circles), animated graphics, networks, and many other kinds helps to learn the configurations and arrangements of many social, economic, or business-related structures. Even abstract concepts such as emotions can be examined by sketching human figures and facial expressions accompanying theoretical analyses. The following chapters comprise descriptions of visual metaphors and visualizations that support knowledge transfer. By developing a habit of sketching objects and events under study, many readers of this book may find a visual side of their imagination and understanding helpful and engaging.

Openness to novelty

Understanding advances in science and technologies requires immediate attention because they are needed in many professions and should be more present in current curricula. Possible answers to this issue require the involvement of young generations of students, described in Chapter 2. Our audience is changing, with generations shifting their interests and attention. With the learning content linked with the reality of current and future jobs, businesses, and occupations, novelty should be presented in an attractive, uncomplicated way with potential for practical applications. Openness to novelty has become essential because technological solutions include multiple new methods invented in various countries. These changes create a need for inclusiveness, better communication internationally, and creating new cooperation opportunities. There is also a need for sharing novelty apprehension and communication with the international scope.

Preparing learners for future careers

Because of the better access to technology, prospective employers expect that students can create visual presentations of data using colors and symbols to better understand concepts under discussion. They look at students as future workforce members and hope they can develop their own unique portfolio confirming their abstract and cognitive thinking abilities. They are often hiring developers according to the stars they received in voting. Many believe that we do better what we enjoy doing. Students know many current technological advances but must become confident in their ability to depict information visually. Preparing students for those challenges can be done in a way they would enjoy.

The integrative method of learning is a tool for preparing learners for their future careers and work. Recruiters expect new employees to be informed about other areas of knowledge and skills, as no specialist can learn and work within the boundaries of only one field.

Collaborative working on projects as a visual learning strategy

Project members learn a lot when working on group projects, for example, when they create computer-animated films/videos. They derive knowledge from creative, technical, and material resources. Project participants scan their sketches or use images of visually inspiring events and processes related to the project, obeying the copyright rules. They learn about visuals: first, the elements of design, which include points, lines, shapes, masses, spaces, motions, lights, colors, and textures; and then the principles of design, which include harmony, unity, balance, hierarchy, and emphasis. By choosing specific color selections, they use additional lines, apply texture, enhance the depth of the picture, or change its balance and composition. They apply basic visualization techniques using color coding, changing proportions, modifying patterns, copying and pasting a selection, and/or introducing symbolic representations. Some use modeling clay to develop three-dimensional models.

Learning projects and activities support the description of tools in the following chapters. Projects are designed to be cooperative, integrative, and visual. They introduce novel trends and technologies. Learners, instructors, and teachers are encouraged to apply strategies described and presented in these learning projects. They hopefully find them useful for their work.

Sketching every day

This book advises the readers to sketch a lot to support developing their visual literacy. The leading advice is: make it your habit to sketch every day something that surrounds you daily – things, events, and people. You can do it using apps installed on the phone or watch. For example, visualize food items through your senses and draw a pictorial list comprising short sketches of groceries you need to buy when shopping. Make quick drawings depicting people (be sure they agree to be sketched) you know or you just met (Figure 1.4 a, b, c). You may focus on their activities, for example, putting quickly on paper a group of jazz musicians with their instruments (Figure 1.5 a, b, c), or people eating in a bar (Figure 1.6 a, b, c, d), or create quick self-portraits displaying your present emotions and involvement at the moment.

To make a sketch, watch, see, and observe what you are sketching (looking and seeing is not the same thing). Thus, developing a habit of sketching is aimed at enriching our perception and understanding.

Figure 1.4 Three short sketches of people from the neighborhood

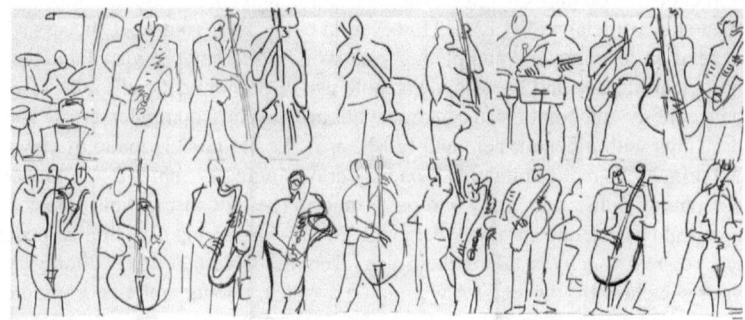

Figure 1.5 Short sketches of the jazz players

Figure 1.6 People eating in a bar. The printmaking, aquatint, aquaforte, and mezzotint techniques

Short activities are interwoven into the text to enrich the book's reading. They encourage readers to make quick sketches related to themes under discussion. Apart from making short sketches, strategies supporting visual learning include using color coding, making visual timelines about the learned issues and events, creating characters and then using them for animations, visual storytelling, and designing serious games about subjects under study.

When you finish drawing or sketching a picture, save it in at least two places, titling it with your name and the title of your project, for example, Your-Name_HistoryofLove.jpg.

Please put a copyright sign on your compositions: to do that, place the type tool (cursor) on the page, hold the option key and the letter g, then write 2023 and your name. Then flatten the image. You may choose to type: Copyright 2023 Your Name.

Guides for working on learning projects

Visual projects are designed to help you understand and memorize the concepts described in the following chapters. It's up to you which tools or media you will choose. You may continue to apply traditional media or use prompts to activate the artificial reality apps. However, in both cases, your final success will depend on planning and completing your work based on your visual and technological literacy and the knowledgeable use of real or artificial tools. To enhance these vital points, you can draw your projects on a computer, create hand drawings with a pencil, pen, and brush, or make 3D models made of clay or 3D printed. To make animation, you may draw single keyframes for animation, use photographs, scan, copy and paste images (being conscientious about the copyright rules), and combine them with your works. Using a traditional pencil lets you rest your eyes, taking them away from the screen. You may photograph your sketches. If you prefer sketching on a watch, phone, tablet, or computer, remember to wear blue-light-protective glasses, even if your vision is perfect.

Color coding. Apply a color coding system to mark categories of your material with different colors. When you practice or describe several categories (such as comics, animation, and video or painting, sculpture, and architecture), assign a different color for each category. The coding will help you and your audience grasp a bigger picture and understand your work. Create a legend explaining the meaning of each color. Follow this legend for various categories in your further work, for example, red for comics, green for animation, and blue for video.

Icon coding. Create icons, one for each category. Choose icons that represent, as symbols, the categories you are working on. This way, people looking at an icon can quickly link the artwork's object with a category it fits into. Coding with icons is helpful because it evokes your subconscious reactions. With the color coding, you have to look at a legend to see which color was chosen for what category. Symbols act as images and allow fast recognition of the category. Icons used on the internet, portable devices, and social media shape our way of living, making it more visual. Many jobs rely on visual methods of presenting information.

Software and apps. Here are some bits of advice about the technology you may apply: when you are working on a computer, you may use professional programs, such as Adobe Photoshop or Adobe Illustrator, or download free apps. You may also download on your computer, phone, or tablet graphic programs such as *(free download): Gimp (www.gimp.org/) (digital/raster), Krita https://krita.org/ (digital/raster), InkScape, (analog/vector) – free download graphic editing software, or the Adobe 30-day trial version, before you decide to buy them. The Gimp and Inkscape are free, as are Krita programs or Adobe Photoshop, Illustrator, or Fireworks.* You may also draw single keyframes for animation, scan, copy and paste images (but be conscientious about the copyright rules), or use photographs and combine them with your works.

If you'd like to program your work, You may download an open-source programming language at Processing.org without charge. MIT alumni Ben Fry and Casey Reas created this award-winning program for anyone willing to try one's exploratory way of programming. This program presents the image, animation, or interactive work along with a code, and the user is welcome to alter the code and see the changes made to the visual data provided by the authors or write their own code following the rules provided in the added manual. Many people avoid coding, but this visual approach to programming offers a library of images, animations, or interactive content for users to start with. Once you modify the code, the content will change. It provides a very intuitive way of learning about coding, which will benefit you on the job by helping you understand and communicate with coders and making you a better prompt writer for AI.

When you work on finishing each project, you may think about it as if it was a report you would send to people who do not speak English. It may resemble the booklet *Point It: Traveller's Language Kit* (Graf, 2009). *Point It* is a set of photographs representing most situations we experience assembled in a pocket book (in English, Spanish, French, Italian, German, and Russian) that one can show abroad to communicate visually with no regard to language or alphabet used there.

When you work on finishing each project, you may think about it as if it was part of a report you would send into space as a visual summary of our knowledge addressed to possible extraterrestrial beings. The message to potential residents of habitable zones on planets or stars in the cosmos would thus comprise a collection of your visual and written projects.

Online technologies support working on learning projects (Harasim, 2017). For your convenience, here is a list of more accessible, open-source programs that typically are not hard to use. If there are any problems, user support groups can be helpful. While some programs need to be downloaded and installed, others are readily available through a web browser; some can sync through mobile devices. Others offer more advanced, paid versions.

2D

- Krita (drawing)
- Gimp (drawing)
- Inkscape (drawing, illustration)

- TV Paint (Animation)
- Autodesk Sketchbook, similar to Adobe Illustrator, is a downloadable 2D drawing program with layers

3D

- Houdini
- Sculptris, which is a free 3D sculpting program.
- Blender is a free 3D creation program that is a bit different then Sculptris
- LibreCAD
- 3D Slash
- SketchUp
- OnShape

Video editing and media

- VLC Media Player
- Handshake
- SMPlayer
- DaVinci RESOLVE

A short orientation for readers

Chapter 2, "Our audience is changing, with recent generations shifting their interests and attention," examines current audiences' characteristics by describing Millennials, Generation Z, and Generation Alpha. Traits and the lifestyles of young people are described, including their attention span, learning habits, quiet quitting, and quiet hiring behavior. The text tells about possible pedagogical responses to changes in the profiles and attitudes of young people, connecting the learning process with specialists' inventions, discussing the teachers' preparation to make the instruction and knowledge transfer adequate for current tasks, and indicating a need to reexamine long-established dichotomies.

Chapter 3, Dedication to sharing, equity, diversity, and inclusion," addresses the needs of the broad spectrum of learners. It discusses selected characteristics of working and communicating with learners: a need for finding shared values, knowing the society, race, and culture issues, getting in touch with learners, caring for their well-being, sharing multicultural values, checking our consciousness, and helping incoming students, among other topics.

Chapter 4, "Thinking and creating; cognitive, creative, and psychological capacities that support collaboration," surveys mental capacities that may enhance visual communication and support creative collaboration, learning, and teaching: artificial intelligence literacy, visual, aural, and technological literacy, cognitive thinking skills, abstract thinking, intuition, visual imagination, creative thinking, metaphorical thinking, semiotics, intelligence and individual intelligences, and spatial abilities.

Chapter 5, "Techniques and strategies in introducing novelty in instruction," discusses strategies for including novelty in instruction and learning. The first part describes concepts and techniques supporting the visual transfer of knowledge: artificial intelligence, AI-based techniques, and other techniques for visually transferring knowledge, including data, information, and knowledge visualization. The second part concerns visual storytelling, serious games, and educational games. The STEAM, STREAM, and STEM programs are discussed, as is the deficit of skilled professionals and technicians in the industry causing the demand for STEM and STEAM graduates. Computer-based learning is another way of introducing novelty in instruction.

Chapter 6, "Collaboration of professionals: old and new examples," describes cooperation and teamwork as necessary when tasks employees perform in most workplaces require input from different specialists. Imagining a successful workforce without a collaborative spirit involving various experts is hard. Collaborative works result in major inventions and successes. There are two reasons for describing inventions. First, discoveries and inventions may involve several areas of interest and competence when they overlap across disciplines and are now interdisciplinary. Second, descriptions of discoveries and inventions create a background for collaborative projects instructors or teachers may want to include in their teaching strategies.

Conclusion

With the learners' experience made more visual, absorbing knowledge and technologies becomes, for many of them, easier and more attractive. Visual shortcuts and aesthetics of solutions that learners find by themselves are grasping their attention. Generating novelty in content, inspiration for new stories, and focusing on equity and inclusion are essential in designing projects.

Progress in technologies has created a demand for schooling students as future professionals and technicians in new areas of knowledge. Along with developing their technological literacy, current workforce members need to learn ways to communicate visually. Education and training in the arts should support the development of abstract thinking and expand the role of computer graphics and coding.

2 Our audience is changing, with recent generations shifting their interests and attention

Introduction

This chapter offers an analysis of a prominent part of the current audience. People born from the 1980s till now comprise the largest generation of contributors and consumers globally. While discussing new ways of communicating, knowledge about changing characteristics of our population is highly important in educational terms, workplaces, and marketing.

Our current audience belongs mostly to generations called Millennials and Generation Z and makes up a significant part of present society and the current workforce; Generation Z and Millennials combined will soon make up 75% of the workforce (Goh, 2019).

A need to know how the audience changes

Beresford Research (2022) defined the age range by generation. Table 2.1 presents the period of birth and the ages of several generations living in the 20th and 21st centuries.

Millennials

The Millennial generation includes people born between the early 1980s and the 1990s. An initial study based on a 2012 analysis of two databases of 9 million high school seniors or entering college students stated that "The trend is more of an emphasis on extrinsic values such as money, fame, and image, and less emphasis on intrinsic values such as self-acceptance, group affiliation, and community" (Main, 2017). However, data and opinions about Millennials changed over time. As described by Main (2017), they are generally regarded as being more open-minded and more supportive of human rights, especially civil rights, equal rights for minorities, and the LGBTQ+ people: lesbians, gays (homosexual males), bisexuals, transgender persons, and queer persons. Other positive adjectives to describe them include confident, self-expressive, liberal, upbeat, and receptive to new ideas and ways of living. The traits ascribed to Millennials

DOI: 10.4324/9781032705347-2

Table 2.1 The years when generation members were born and their age range are based mainly on Beresford Research (2022).

Generation	Born	Age
Generation Alpha	2013–2025	10–(now)
Generation Z	1997–2012	11–26
Millennials	1981–1996	27–43
Generation X	1965–1980	43–58
Boomers II	1955–1964	59–68
Boomers I	1946–1954	69–77
Postwar, Silent generation	1928–1945	78–95
The Greatest Generation*	1901–1927	96–104

* *This generation's name refers to a book,* The Greatest Generation, *written by Tom Brokaw (2001), a journalist for NBC.*

include their tendency to value meaningful motivation. They can challenge the existing state of opinions and ideas their superiors proclaim but cherish good relationships with them. They have better knowledge of technology and are open and adaptive to change and to learning new things. Many are free-thinking and creative and value teamwork and social interactions in the workplace (Cheng, 2019; Main, 2017). Millennials have often been depicted negatively, but conversely, authors also describe them as one of the most adaptive and creative generations (Indeed Editorial Team, 2022). The Pew Research Center (Funk, 2021) found them to be more concerned with intrinsic and moral values over extrinsic and material ideologies. They have more racial and ethnic diversity, better education (36% of men and 43% of women have a bachelor's degree or higher, with better income when they have a college education), and greater participation of women in the workforce (Morrison-Williams, 2022). They marry later (in 2019, 46% of Millennials 25 to 37 years old were married) and live with their parents longer than previous generations (Bialik & Fry, 2019).

Generation Z

The child population in the US is decreasing in size while its diversity is increasing. In 1960, a share of the total population in the US was 35.7%; it fell to 22.1 % in 2020 (O'Hare & Mayol-Garcia, 2023). Generation Z makes up about 40% of all consumers (Patel, 2017) and is a growing percentage of the workforce. Generation Z is the largest generation on Earth, accounting for 2.47 billion (32%) of the 7.7 billion inhabitants of Earth, more than the Millennial population of 2.43 billion (Spitznagel, 2020).

According to the Pew Research Center (Barroso, 2020), Generation Z members tend to be racially and ethnically diverse, progressive, and progovernment. About 29% have a bachelor's degree or higher. With limited access to technology, the low-income members of Generation Z are vulnerable when they enter the workforce. A slight majority of Gen Z-ers (52%) is white; 25% is Hispanic, 14% is Black, and 4% is Asian. According to the Annie E. Casey Foundation (2021) survey results, "For many Gen Z-ers, the backdrop of their early years included the country's first president of different ethnicity and the legalization of gay marriage. They are more likely to have grown up amid diverse family structures – whether in a single parent household, a multi-racial household, or a household with blurred gender roles. As a result, they are less fazed than previous generations by differences in race, sexual orientation or religion."

Generation Z members are seen as pragmatic consumers, interested in exploring and evaluating many options in terms of their extensive social networks and recommendations of real-life users rather than celebrities (Broadbend et al., 2017). Also, they often shop online and choose things that express their values and identity: they prefer to buy sustainable and personalized products and brands, choosing those that share their political profile (Annie E. Casey Foundation, 2021). Many plan to settle in a country house in a rural environment to live outdoors (Figure 2.1), according to the cottagecore trend (BBC The Collection, 2020). Generation Z has a lot of time yet to grow and mature, so surely, successive surveys will provide new data.

Candidates for hiring belonging to Generation Z complain that their education has not prepared them to enter the workforce. Gen Z has not developed "soft skills, such as negotiating, networking, speaking confidently in front of crowds, and developing the social stamina and attentiveness required to work long hours in an in-person environment" (Gartner, 2023).

Figure 2.1 Boats; a linocut

Generation Alpha

The Generation Alpha group will be diverse concerning race, ethnicity, family structure, and finances. The oldest ones, now 9 or 10 years old, are often the children of Millennials (1981–1996) and the younger siblings of Generation Z members (1997–2012). Individuals born between 2013 and 2025 share common statistical characteristics. These individuals, predicted to reach 2 billion globally, are almost entirely born in the 21st century, so they are digital natives immersed in technology. They are growing up logged on, linked up, and tutored by Siri, Alexa, various visual media, text-to-speech translators, artificial reality–based systems, and more apps and devices daily. About 90% of children used a handheld electronic device by age 1; in some cases, children started using them when they were only a few months old (LaMotte, 2019). John S. Hutton uses magnetic resonance imaging (MRI) to research the impact of reading versus screen use by kids. He observes that about 90% of them use screens by age 1. He conducted studies on kids 2 to 3 months old using screens (Hutton et al., 2022).

Members of Generation Alpha still need to join the audience of this book or any other one except children's books. However, they deserve full attention now because many of them are already using smartphones and tablets as their toys and devices, enhancing their digital literacy and adaptability. Electronic technology, social networks, and streaming services define these kids' entertainment, while traditional television is much less important. Actions their families and teachers perform now will indeed define their future in a better or not-so-good way.

Personality traits and the lifestyles of young people

Attention span

The theme of an attention span, defined as the length of time one can concentrate mentally on a specific activity, became widely discussed after Microsoft performed a study on the attention span in 2015. Daniluk et al. (2017) offered a neural language model with a key-value attention mechanism, while the Statistic Brain Research Institute studied attention span statistics (Statistic Brain, 2018). The average attention span in 2000 was 12 sec, while in 2015, it was 8.25 seconds. Allegedly, Millennials in the US typically have an attention span of 12 seconds, while Generation Z has an 8-second attention span, shorter than that of a goldfish, which is 9 seconds (Tongwaranan, 2019; Ryssdal, 2014). Our audiences are now constantly looking for new solutions, inventions, brand-new stories, and surprises. However, according to Zauderer (2022), childhood development experts say that the average attention span in a child, when they can stay focused on a particular task, grows 2 to 3 minutes per year of their age. The average teenager who is 14 years old has an attention span of 28 to 42 minutes, and the average 16-year-old has an attention span of 32 to 48 minutes (Cross River Therapy, 2022).

Several factors may shape reasons for the shortening of the attention span. Millennials and Generation Z are primarily digital natives (Generation Z are the first true digital natives), and many are technologically fluent (Thomas, 2011). They can enjoy fast internet speeds, allowing access to an ever-changing and thus endless stream of data, better availability of digital media, and the world accessible at their fingertips, e.g., via pocket-sized portable devices, and multiple choices for communication.

Possible benefits of the shortened attention span

While students focus on a single activity for a shorter time than before, the shortening of attention also provides benefits. Generation Z people call it an 8-second filter, which is 10 times more effective than in the Millennial generation (Patel, 2017). It helps them quickly assess whether the information presented is relevant and enables them to focus. Attention is becoming more intensive and efficient (Goh, 2019). Students can do multitasking better than before. Millennials can scatter their attention. They bounce between three screens intermittently, using applications such as Vine and Vine Camera, Snapchat, and TikTok. Generation Z can watch five screens simultaneously: a smartphone, TV, laptop, desktop, and tablet used to click on a blog post, watch a video, or view an Instagram photo.

Physical and mental well-being

Many authors worry about the mental health–related problems that might result from the lifestyles of the Millennials, Generation Z, and Generation Alpha members. These generations spend more time on digital media than reading (Sliwa, 2018). A group called the tweens (8–12 years old) spends 6 hours daily on media use, and the teens group (13–18 years old) has 9 hours of average daily media use (Common Sense Media 2023). After researching the current situation, the World Health Organization and the United Nations health agencies have issued guidance about the screen time used by children. They both advise that children under 5 should have 1 hour daily or less of watching screen and no screen time for babies under 1 year (AP News, 2019). The activity that follows invites you to transfer your attention briefly from a screen to a natural milieu (Figure 2.2).

Activity: Transposition

Sit under a broad-leaved tree: oak, birch, elm, walnut, beech, poplar, maple, willow, or of any kind you find not far from you. Listen to the leaves, even

Figure 2.2 Landscape sketches

if they are motionless. Convert this experience to a music composition by clapping your hands against a soft surface. Then, draw short sketches of the trees and leafy bushes, drawing from observation, your memory, or fantasy. How can you connect music with the sketches?

Researchers, worried by the fact that screen time use by infants, toddlers, and preschoolers has exploded over the last decade, scanned the brains of children 3 to 5 years old and found lower levels of development in the brain's white matter – an area key to the development of language, literacy, and cognitive skills (LaMotte, 2019). In another study (Hutton et al., 2022), researchers performed MRI and cognitive testing, revealing changes in cortical thickness and other structures across the cerebrum, supporting visual and higher-order processing skills. Studies show that allergies, obesity, and other health problems, including mental health among children, related to the use of digital media with a screen (for example, a smartphone, computer, video game console, or television) have become increasingly prevalent (Annie E. Casey Foundation, 2020a).

Numerous studies indicate that long screen time affects about 50% of computer users. They suffer because of dry and irritated eyes, along with blurred vision. A prolonged focus on a screen may result in asthenopia – eye fatigue, which means eye discomfort, dimness of vision, and headache. Eye health professionals suggest the 20–20–20 rule; every 20 minutes, one should go off the screen for about 20 seconds and look at a distant object around 20 feet away. They also recommend using blue- and ultraviolet-light-blocking glasses, plain or with prescription. Some parents use blocking apps that limit their child's screen time on their phones, tablets, and other internet-connected devices.

Members of recent generations report their mental health as fair or poor: 27% of Generation Z members, 15% of Millennials, and 13% of Generation X members, according to a survey conducted on behalf of the American Psychological Association (after Annie E. Casey Foundation, 2020b). Many complain they experience a state of dissatisfaction with life and dysphoria (the opposite

Figure 2.3 Two sketches of the Generation Z members engaged in a social gathering

of euphoria). America's younger generations have also received more mental health treatment than previous generations: 37% of Generation Z and 35% of Millennials have reported doing so compared to just 26% of Generation X members and 22% of baby boomers. Too much screen time may evoke feelings of isolation and result in underdeveloped social skills (Annie E. Casey Foundation, 2020b). Many individuals feel frustrated, unconfident, or disoriented (The Economist, 2019). Generation Z members are prone to participate in protests of different kinds and treat their projects as manifestations of their objections. But it's not just Gen Z – everyone's social skills have eroded since 2020. Gartner (2023), a technological research and consulting firm, indicates that employees' burnout, exhaustion, and career insecurity, heightened during the pandemic, now negatively impact performance. He stresses a need to focus on employees' mental health and to snag in-demand talent. Figure 2.3 presents two sketches of Generation Z members engaged in a social gathering.

Activity: Power

Make a list of five strengths you possess. Think why people you know like or appreciate you.

For example, you may have strengths such as being a leader, fast learner, organized, playful, having a sense of humor (don't say that you do not possess it).

Learning and working habits

In formal classroom learning guided by a teacher, multiscreening reduces the ability to focus on one repetitive and boring task. Reading is considered tedious; movie watching is linear. In informal, self-directed learning, attention lasts as long as the topic is of interest. Work performance becomes the priority, often combined with personal commitments. Corporate learning programs can be formal or informal. Following a corporate learning program requires self-control. With intensive and efficient attention, it is easier to determine whether the information is relevant and to include learning in one's daily workflow (Goh, 2019).

In many cases, students from Generation Z are changing their learning and working habits. They are no longer passive as content consumers; they try to modify and upgrade the offers available on the online market. Depending on the type of material they tend to learn online. One of Generation Z's most talented and successful members is Gitanjali Rao, born in 2005. Recognized as America's Top Young Scientist, Rao is an inventor of Tethys – an early lead detection tool (Prisco, 2018), Epione for early diagnosis, using genetic engineering, of prescription opioid addiction, and Kindly – an anticyberbullying service using AI and Natural Language Processing (UNICEF Office of Innovation, 2020). Rao, who received the EPA Presidential Award, is a STEM promoter (Rao, 2021) and social activist (Secher, 2022).

Quiet quitting

Some people are satisfied by minimal accomplishments, such as mere participation, for example, in cooperative projects or competitive sports (Turner, 2023). Many employers think they have unrealistic expectations of working life. The strategy of quiet quitting taken by many employees may result from recent political, economic, and social events, which may result in poor corporate culture, low pay, lack of opportunities for advancement, feeling disrespected at work, having childcare issues, lack of flexible hours, and not having good benefits. Employees feel a need for recognition and a need for clear direction. As they stated in online blogs, quiet quitting, a trend started on TikTok, involves completing one's work responsibilities without going above and beyond, such as logging out at 5 p.m., not seeking additional tasks or projects, and taking regular time off (Bretous, 2022). As a response, an employee is only meeting the minimum performance requirements; does not pursue a promotion or seek any praise from their manager; isolates from coworkers; refrains from company activities; and shows burnout (*www.globalization-partners.com/blog/4-signs-of-quiet-quitting/#gref*). According to tweets, in many cases, a boss wants this person to quit, so this employee does not get new or challenging assignments anymore, has their benefits or job title changed, and does not receive support for professional growth. The boss avoids this person, who becomes excluded from meetings and conversations.

There is much interest in how quiet quitting can be stopped and replaced with another attitude. Within the quiet quitting internal reference point, people still meet their job expectations but prioritize other aspects of life (Shankar,

2022). They determine how they want to spend their time and evaluate how much time they can give to work and how much remains for personal lives, family, friends, and communities. Thus, a boss or manager must accept that while employees are getting their work done, they protect their private time within boundaries. They should reframe a strategy for employee productivity, set clear and reasonable expectations, explain the company's mission or purpose to tie employees to the organization, and explore when individuals feel most satisfied, fulfilled, and rewarded (Shankar, 2022).

Quiet hiring

Quiet hiring is a trend developed by employers that offers new ways to find in-demand talents. Employers tend to acquire new skills and capabilities of employees without adding new full-time employees. They focus on internal talent mobility, enhance the upskilling opportunities for existing employees, and focus on alumni networks, gig workers, and nontraditional candidates. Quiet hiring means enhancing employees' skills to fill the skill gaps and giving them more responsibilities. It includes outsourcing some projects to freelancers or contractors if they are less expensive or faster. However, this does not secure employees' development. Some projects, previously solved by outside talents, are now done in-house. Investing in employees takes place when an employee has the chance to develop and apply skills that were absent from their job description. Employers move some employees to advanced positions with more responsibilities and better salaries.

Possible pedagogical responses to changes in profiles and attitudes of young people

Teaching, assigning tasks, and advertisements' content must now relate to real-life situations. While they are constantly logging on and linking up, even in a classroom or workplace, members of recent generations stay in their coordinate systems, defining their thinking and the reference points that fix their evaluation of what they encounter. Many of them lose interest in what they are listening to or reading when they find the content unrelated to real-life issues or indirectly connected with their current or future business life. For this reason, these people constantly seek confirmation that the content of a lecture, their currently performed task, or the ads they watch fit and agree with their action plan. They use their phones to connect with other social group members to discuss what they hear or read.

Because of their sensible, realistic, matter-of-fact attitude, some individuals consider beyond-college learning, graduate and postgraduate studies (such as a master's degree or a doctorate) a waste of their time and money. While others pursue academic careers to teach or conduct research studies, more individuals need time and money to realize their practical plans.

In the workplace, many young members of the workforce can comfortably perform their tasks, such as counting as a bank teller or organizing objects, and simultaneously talk to their coworkers, who must focus on their work without any disturbance, so they become irritated when they lose their count (Rothman, 2023).

The corporate world needs new strategies to grasp students' and employees' interest in learning or working. For many, only contemporary events and issues matter; the past is unimportant. No interest in the past results from a strong determination that only current links with reality are essential for practical reasons. Moreover, many members of the new generations are uninterested in scientific, social, and political matters because they believe they can do nothing about them.

Building motivation to learn

With all factors in mind, new, engaging projects are needed with which teachers can motivate current generations of students. Many times, individual students who feel anxious, depressed, or bored, when encouraged to start working on a theme that they find personally interesting, change their attitude and can focus on learning. Often, borrowing, copying, rephrasing, or mimicking somebody else's work inspires students' projects. It would help if retelling old tales and stories and reusing the characters were replaced with action-based stories specially designed for the audience.

Building a need for serious studying is difficult with these attitudes. With an AI-based app or Google, one can obtain minimal information and find an answer that is enough to satisfy one's needs. There are many summaries of fiction and nonfiction books and videos, and they are advertised as a source of learning the key points, getting insight, and understanding ideas and facts. Many students find such abstracts or overviews satisfactory. They do not return to the masterpieces of literature or the most informative nonfiction books because it has become trendy to hate reading. Those studying world literature often use plot summaries and the so-called in-depth study guides to avoid reading the whole book. AI-based apps also summarize academic articles, magazine articles, business reports, news, blog posts, and more. Previously, the use of such aids was prohibited.

Actions taken by various organizations aim at accelerating learning. This trend has repercussions in school and academic curricula. Despite the values of learning about the Renaissance, Baroque, and other styles, which result in coming into contact with beauty merged with scientific thinking, some art history courses are now replaced with courses in ethnic or women's studies to provide students with an opportunity to discuss their current individual problems. According to another trend, some design assignments advise students to rework the solutions that already exist online instead of creating their own projects. Concepts like talent, creativity, thoughtfulness, formulas, abstract thinking, cognitive thinking, and innovative design are not common in all those instructional strategies. With opportunities offered by AI-based apps, what would we call thinking enhanced by their output?

Very often, individual students or workers ask why they must listen to a long lecture and memorize useless facts and numbers. They wonder why they do things that a computer may do without their intellectual and time-consuming involvement. A physics or chemistry teacher should obtain materials to answer these questions by providing real-life examples and practical applications of theories and laws. Students can then discuss the advantages of applying knowledge to their daily activities, jobs, or future professions. They can also learn to value informed cooperation with others. When thinking and creating, examples of scientific and technical solutions help connect conceptual and concrete ways of thinking. Projects in Chapter 5 of this book unify separate things and concepts while telling about natural processes and events. AI as a personal assistant may become helpful here.

Connecting the learning process with successful results of specialists in different areas

To engage individuals in active participation in learning and working, it seems necessary to show them how the content they learn may connect with their current or future professional work and practical actions. There is a need to encourage everybody to think about how they can apply the subjects of their current task to their future practice. Various forms of art, music, sciences, mathematics, business, education, or communication media may seem irrelevant to young people's lives. They should consider these themes as their possible tools of trade. The aim is to convince these individuals that it takes a creative mind to be good at other fields of study and branches of knowledge, to see the spaces between them, and to find all the cool things that come out of them. For this reason, Chapter 6 of this book offers ready examples of past and present collaborative inventions and teamwork by professionals that changed the practical life conditions, to support teachers, instructors, and employers.

AI helps make things as good as possible, from spelling, syntax, punctuation, perspective, proportion, or rendering to quality texts and attractive products. We need to focus on using these outputs to create new, challenging, beautiful, yet thought-provoking and inspiring content. While AI can do the research, write a story, illustrate it, then convert it to voices, videos, and more, learners should make knowledgeable, aesthetic-based selections and make them attractive through collaborative efforts. Having more time and the help of AI assistants, they can become more creative, curious, persistent, observant, knowledgeable, and social to create quality products that find their market, can be appreciated, and build cultural values.

Preparation of teachers to make the instruction adequate for current aims

Many teachers need some help in gaining novelty comprehension and require support in instructing technologically literate generations (Millennials,

Generation Z, and Generation Alpha). They should get help in providing young people with integrative, interactive, globally current, and easily applicable materials. Connecting students with technology in a meaningful way requires a teacher's creativity. Instructors need materials that grasp learners' attention. This need is caused not only by the shortening of their attention span. Members of recent generations, both students and people working for their own or other companies, use their 8-second filter to extract the essence from the lecture's content, and they do it according to their frame of reference: Will that be useful in their career, and will it support their future financial gain?

In many cases, we cannot interpret and comprehend real encounters with members of recent generations in traditional paradigms. They modify cognitive actions to fit their newly acquired abilities of applying an 8-second filter and dividing their attention between many topics of interest. Social researcher Mark McCrindle (2022) stated his pedagogical aims: "Visual communication is increasingly important in the digital age. Utilise the power of visual design to communicate data with clarity, articulate complex ideas, and break data down into digestible information. We take complex information and crystallise it into simple stories so you can make faster, more confident decisions."

Reconsidering the long-established dichotomies

Conventional, widely accepted dichotomies are dividing and contrasting some cognitive and psychological traits and activities as if they were opposed and different. Current information about characteristics typical of members of recent generations may create a need to reexamine concepts such as left- vs. right-hemisphere functions; visual vs. verbal ways of thinking, learning, and retrieving memories; introvert vs. extravert people (Cain, 2013); convergent concrete thinking vs. abstract divergent thinking; analytical vs. creative thinking; sequential vs. holistic thinking; intrinsic vs. extravert motivation and behavior; and more. For example, in *The New Yorker*, Joshua Rothman examines the thought process, stating that "Thinking in pictures, thinking in patterns, thinking in words – these are quite different experiences. . .. Visual thinkers and verbal thinkers may represent points on a continuum" (Rothman, 2023, p. 27).

Preventing or diminishing the effects of long screen time

We need several kinds of actions. Students and employees may be encouraged to make quick sketches of objects and situations they are reading or learning about. The habit of sketching may enrich learning with an ability to visualize things and ideas. At the same time, it may avert the eyes from the screen for a short time and prevent the development of eye fatigue symptoms.

Students may download a free drawing or sketching app. Sketching on paper could be encouraged because it takes their eyes away from the screen's radiation.

Students may also scan and make screenshots of pages containing long texts by pressing (on the Macintosh computer) and holding simultaneously three keys together: 'Command,' 'Shift,' and '#3' for a screenshot of the whole screen, or 'Command,' 'Shift,' and '#4' to select a part of it; it will generate a. png, a portable network graphics file named Screen Shot, followed by the date and time they took it, saved on a desktop. For example, "Screen Shot 2023-04-26 at 4.50.55 PM." On a PC, there is a print screen button, and pasting the content to any image-editing software allows for preserving and modifying it. After saving a file, one can work on the project anytime. Students may also store data as QR codes (quick response codes) and graphical representations of digital data. They may later scan a QR code on a camera found on a smartphone or other device.

Conclusion

According to the data and reports about new learners, making teaching and instruction strategies suitable in schools, academies, businesses, and marketing companies seems urgent. We should entertain more, intensify fast and compressed knowledge delivery, and apply gamification more, with rules of play, point scoring, and competition with others adapted for learning. Collaborative activities in project production are helpful so the learners can choose their tasks and learn through discussing and sharing. It can make' every team member comfortable; thus, creating a new project will be fun and bring new ideas and ways of creativity and thinking. Also, recreating and organizing knowledge, concept mapping, creating art about new concepts, and building maps, timelines, and models can help the participants visualize knowledge. Further chapters offer some inquiry, and the integrative visual learning projects aimed at knowledge transfer make it easier for the learners.

This analysis means that the book's focus on who gets rewarded on the job – creative, abstract thinkers, collaborators, and strategic minds – puts forward conclusions: to return to the values of chess playing and learning the basics and logic behind different languages. For example, learning Latin may help those who use different alphabets and characters, while learning the Asian or Arabic languages may serve those who live within Latin-based cultures well.

3 Dedication to equity, diversity, and inclusion

Introduction

Our society needs to find and share common values. Most members of a cultural environment in an individual country adhere to its accepted values, thought systems, and multiple viewpoints. However, in some places, books are being destroyed or excluded from the curricular practices. With all the diversity of world cultures, creating safety and success in the learning environment requires securing inclusiveness in knowledge delivery. Students, employees, and other readers may ameliorate their attitudes and improve the learning process when working in the pervading mood of goodwill, friendliness, cooperative feelings, and acceptance of all groups and parties. Hence come the requirements concerning cultural values that involve instruction, such as equity, inclusion, and multicultural sharing. A company needs similar values for successful communication, collaboration, and cooperation. These needs are even more severe and urgent regarding existing conflicts and epidemics.

Studies on society, race, and culture

Controversies on teaching about wokeism resulted in a revival of interest in critical race theory. Currently, 'woke' means to be informed, educated on, and aware of social injustices. Wokeism promotes sensitivity to systemic injustices, prejudices, and tolerance in our country and worldwide. Critical race theory, an academic movement regarding society, race, and culture, proposes that any analysis of American society must consider its history of racism and how race has shaped attitudes and institutions (Borter, 2021). 'Critical' refers to critical thinking, critical theory, and scholarly criticism. "Critical race theory (CRT) movement is a collection of activists and scholars studying and transforming the relationship among race, racism, and power. It examines the foundations of the liberal order, including equality theory, legal reasoning, Enlightenment rationalism, and neutral principles of constitutional law" (Delgado & Stefancic, 2023, Introduction). The book *Critical Race Theory: Critical America* is a collection of essays, a cross-disciplinary study of how laws, social and political movements, and media formulate social conceptions of race and ethnicity.

DOI: 10.4324/9781032705347-3

Getting in touch with learners

To share knowledge and effective instruction, an instructor, teacher, or business coach must be well-understood and prepare learners to participate actively, which benefits them. This kind of instruction means looking from a student's perspective, analyzing one's skills, strengths, talents, and affinities. Howard Gardner's theory of multiple intelligences (Gardner, 1993/2006) proposed that each person may possess various forms of intelligence but always has a dominant one. Learners find out for themselves who they are, which is their strongest intelligence, but often, they need the help and understanding of their teachers and instructors.

The essential tasks for teachers and instructors involve delivering novel ideas, technical terms, and complex or abstract concepts in the most precise, most understandable way. One has to make good contact with those identified learners who are unlike across various social, ethnic, or generational environments. A class or workforce of great diversity may include groups with diverse backgrounds, skills, modes of perception, and learning styles. With dramatic events going on in many countries, for the first time in their lives, kids saw pure evil and pure good, as someone said. Supporting their happiness, peace of mind, and success becomes a crucial moral obligation. Getting in touch and starting a convincing, helpful connection with people who may feel isolated and under strain may require building individual contact and giving a personal touch in instruction.

Looking after the well-being of learners

In many learning and working environments, equity and impartial instruction aim to care for the minds, not only the learners' bodies. They address the fragile points in their mental health. As discussed in a previous chapter, many learners declare their barely fair or poor state of mental health and a need for help or treatment. Moreover, many learners experience feelings of isolation and underdeveloped social skills, probably because they have been spending too much time isolated at home in front of screens of different types (Figure 3.1).

While designing cooperative projects, exercises, and discussions, instructors and teachers can considerably help anxious individuals. For example, distressed students can be given individualized projects based on carefully analyzing their conduct. In small or large workforces, a supervisor or manager can evoke interest in an indolent employee by assigning motivating tasks.

Activity: Finding the balance

List and describe five good qualities in a person you do not like.

For example, such a person can be sensitive, open-minded, or seen in places that support good actions. You may still not like that person, but you can give them justice. Thus, you will also grow as a human being.

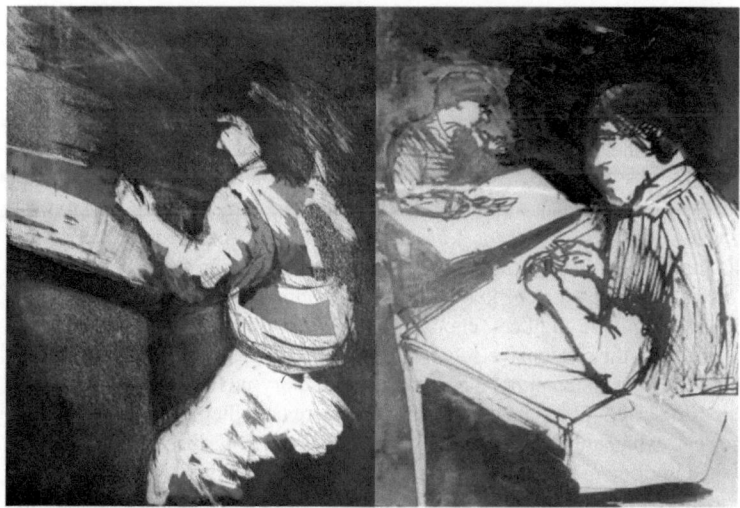

Figure 3.1 Some learners may feel isolated. Images are done in metal printmaking techniques

Also, find three good qualities in a portion of food you do not like and describe them.

For example, this food may have color that stimulates your thinking and looks good in your favorite food. It may remind you of something you enjoyed. With a positive attitude, one can appreciate more what is around them.

Multicultural sharing

We can help students and workers coming from abroad and seen by some individuals as belonging to the not-so-good social or cultural sphere. One cannot blame young individuals for doing how their parents and grandparents raised them, and then they were primed by teachers and neighbors. The task of making incoming employees and students, who often find themselves lost and unprepared, feel valued and able to be successful is essential. It can be done by arranging open discussions in which learners can speak freely, for as long as they like, about the topic of their choice without constraints. Designing projects related to the cultural values observed in the individual learner's country can create an opportunity to present cultural objects and create art projects embedded in their own culture, to know and understand varied assessments about what is beautiful to whom and why such evaluations differ. This action amplifies students' or employees' self-confidence and builds a foundation for the feeling of acceptance by others.

Metaphors for visual communication

Visual metaphors that symbolically represent other, often abstract concepts are suitable and widely used for instruction about new material. However, there may be no use in offering metaphors drawn from ancient or prehistoric times, even when till now they were considered well known and influential. Telling about a Pyrrhic victory (of a king Pyrrhus ruling Epirus in current Greece in the 4th century BC) can be meaningless for many Millennials who are focused on actuality going here and now, contributing to their careers. Also, the labors of Sisyphus (father of Odysseus, king of ancient Corinth in current Greece, who, as a punishment lasting for eternity, had to roll a boulder up a hill every time it rolled back) may not be known to some members of Generation Z, who are seen as pragmatic thinkers, choosing possible options according to opinions delivered by their extensive social networks. Many instructors change the setting and choose metaphors based on contemporary novels. Choosing persons acting in current plays and movies may be more convincing for creating metaphorical explanations. Characters chosen or created by the learners, such as those described in the next chapter, may serve well for conveying metaphorical descriptions of complex concepts.

Checking one's sensitivity, assumptions, and behaviors

The 'special care' issue may present other challenges for instructors, trainers, and teachers. For example, a disabled individual in a wheelchair may be a gifted and talented person, even if not so social, but still bright and open-minded. Any bias, discrimination, or prejudice about disabled people is called ableism. This attitude can cause aggressive laws and policies, a lack of civility and respect, or violent behavior such as bullying or abuse. More often, benevolent ableism is unconscious, such as opening access to some places or benefits for disabled persons and treating it as a 'privilege,' discrimination in favor of nondisabled people, or naming them 'normal' by contrast with the disabled ones. An attitude toward disabled learners draws from a widespread stereotype that disabled people are fragile, weak, and vulnerable and one must come to their rescue. In a learning environment such as a workplace or a classroom, especially in a collaborative setting, it is essential to select the kinds of activities that are available to an individual learner with a specific disability. It helps when this learner lets the teacher or instructor know about their strong aptitudes, skills, and interests and then can contribute with them to a cooperative project.

Jealousy and envy can bring adverse, often uncontrollable reactions, and one may not realize one's motivation. They can be seen in students and teachers. Building competitiveness in a team may bring about good or awful results when a team member is comparing themself with others, having to be better and the most important in any task, gathering, or other situation. Showing up where one is not invited and taking over is an example.

The idea of norms and standards may make envious individuals feel secure when they expect all to look the same. According to expectations, everyone behaves similarly so that nobody will feel envious. However, breaking traditions for some creative outlets can be beneficial. Sometimes, breaking standards for somebody skilled or talented and exceeding norms can be good. There is also a need to break the tradition of putting the gifted and talented learners in one category with those that are disabled or slow, despite the contrasting needs of the members belonging to those two groups. The 'special care' needed for each group is different, even when some students who exceed the standard level of knowledge and thus are bored may cause some trouble to teachers.

Knowing the place well and selecting the proper time to meet learners supports their well-being and may affect the instructor's or teacher's role. It is also worthwhile to know the expectations from a group, as the group styles may vary considerably, being fast or slow, realistic or abstract, funny or business oriented, and wanting to get much in return.

Some current students see art as a way to protest, not always clearly realizing what they are protesting. Sometimes, the reason for an action is less important than the fact of protesting. Readiness for such actions creates the possibility to awaken learners' curiosity and openness to things and actions that may be good for the plan. For example, the organization People for the Ethical Treatment of Animals (PETA) has been struggling since 1980 with the use of animal fur. Now, this movement fights using wool, leather, and down and promotes down-alternative jackets (Kent, 2023).

Looking at our consciousness

Enhancing the visual aspects of our perception, learning, activities, and other forms of existence may bring about valuable experiences. For example, stargazing, bird-watching, and art recognition and appreciation could benefit our well-being and motivational lines of action. Until recently, scientists ignored the human side of stargazing. Now, the contact with the cosmos, its aesthetic values, and the awe it evokes are considered beneficial for our mental health. Marchant describes awe as a "feeling we get when confronted with something vast that transcends our normal frame of reference and that we struggle to understand" (Marchant, 2020, p. 290). Cosmonauts and astronauts returning to Earth often tell first about the awe they experienced in the cosmos while looking down on our planet before talking about factual, scientific issues. Researchers have found that looking at awe-stimulating images improves habitual thinking, making people more interested, happier, less stressed, and more creative. With AI taking care of our hard-to-do tasks, we have time to create awe. The question is how and how to support it. Thus education should go from memorizing, refining, and fact-checking to creating, making connections, collaborating, and independent abstract thinking.

While many believe that everything around us is physical, our consciousness, even though called bodiless, insubstantial, ephemeral, or implicit, is real. We must remember that it is fragile. Working on improving our motives, openmindedness, good intentions, and views may be supported by visualizing such positive values as productivity, respect, and fairness and by respecting the individuality of peers. This improvement may be made by sketching scenes of everyday experiences and the valued or admired objects providing aesthetic pleasure. Developing a habit of watching and recording surrounding objects and actual themes for our tasks may become a good factor in building peace of mind, openness, and flexibility.

Activity: Now and in time to come

Write a message you will put into a bottle. Find a place to hide it for five years to read it after this time. Keep your ideas yours: What do you value most in yourself? In your friend? What do you wish for? Do you expect this message would be different if written after five years? For example, you may insert a message, "Do not compare yourself with others." Alternatively, "Find your strengths, interests, and passions, and follow them to grow." You may decide to put both messages in the bottle.

Examples of worldwide initiatives

Many large international organizations, corporations, and institutions are working on supporting inclusiveness, multicultural sharing, and cultural values. For example:

The Human Brain Project (2023), which involves the work of over 750 scientists from more than 20 countries on neuroscience, computing, and brain-related medicine, is arranging a 2023 event titled We Are Science: Towards a Realistic Future to Enhance Inclusion, Gender, Equality, and Diversity in Science.

The W3C Web Accessibility Initiative (WAI) (2023) organized several symposia, documentation, guidance, and new standards to help people with disabilities. These actions comprise the AI and Accessibility Symposium (*www. w3.org/WAI/news/2022-12-15/symposium/*).

Documenting Additional Guidance for People with Cognitive Disabilities, Low Vision, and Mobile Devices (*www.w3.org/WAI/cognitive/*).

Cognitive Accessibility addresses the needs of people with cognitive and learning disabilities, with current topics about mental health (*www.w3.org/WAI/WCAG2/supplemental/*).

Low Vision Accessibility, supplemental guidance (*AG WG, Low Vision TF*).

Mobile Accessibility supports standards development and exploration (*Mobile Accessibility at W3C. AG WG, Mobile TF)* and more.

An old World Wide Web Consortium, started a long time ago as an informal organization, made sure, from the beginning of the www, that everyone on the planet would see the content of a page the same way, despite of age, price, or type of device and the place from which they look at it.

Nicolas Negroponte made an XOXO action to ensure each child has a laptop – a person would buy one and pay for two: one for oneself and one for a child somewhere in the world. It used Linux as an "unbreakable system" to establish stable connections no matter where the child lives.

Helping incoming students

Colleges and universities are now gaining more students speaking various languages. It happens at the sessions for prospective students that some of them do not speak English at all. Sometimes, students invite an instructor to their graduation parties where parents do not know English. Those students do not experience language immersion or support from home; their life often revolves around communities of people for whom English is a second language, or even not. However, many schools around the globe teach in English. For example, graduate students use the English language to pass their final exams. Coding is also in English.

Whenever a student has a hearing problem, universities arrange for two translators because the terms used in technology-based courses are often abstract and less known to the general audience, and sometimes a translator who is not a specialist. Often, an opportunity to solve problems visually supports students' understanding and retention.

Activity suggested for teachers and instructors of a workforce: Now and before

As a distraction from tiring tasks, ask the learners to carry out an activity in line with the following guidelines:

Refresh your memory and return to your old art projects and drawings from childhood. Draw/sketch from memory some images that evoke your past cultural traits, motifs, traditions, or conventions. Then, please share with others your drawings or sketches and show the designs that were popular in your home and its surroundings where you lived as a child. Ask others to share their pictures or photographs from the past. List at least five features, designs, or traditions that resemble the ones that just came to your mind. Some singing or playing would follow this activity.

Building one's multicultural values

Values accepted in cultures are often driven by religious inspirations, tribalism, sectarianism, caste, racism, clan and lineage, nationalism, political partisanship (Nalven, 2023), or the notions of heroism. However, a hero for one side is an enemy for another one. Zorro, Spanish for 'fox,' a fictional character fighting against corrupt and tyrannical officials and villains, had a high bounty on his head. He was named a criminal in one student's paper and a hero in many others.

To evaluate one's extent of multiculturalism, one has to examine one's values, beliefs, and assumptions regarding one's knowledge of more than one culture. Some employees, students, and teachers have parents coming from different countries. One may ponder whether one can identify with more than one culture and whether or not one has internalized and acquired knowledge of more than one culture (Fitzsimmons et al., 2019). In their research on diversity, Taras et al. (2021) consider diversity a benefit and a challenge to virtual teams, especially global ones. They recognize two facets of diversity, personal and contextual, and find that the contextual diversity of participants adds creativity, decision-making, and problem-solving and enables higher-quality, more creative, and innovative consulting reports. René Lacerte, a founder and CEO of business software company bill.com, which provides cloud-based software, wrote that leadership is not about telling people what to do. It is about "setting up an environment for them to do the best work of their lives" (bill.com, 2022).

Activity: Doing the best work while using free time for enjoyment

Write an answer in four sentences: What would you want to do if you had no duties from now till the end of the day?

For example, I would walk my dog, play chess with my neighbor, and feed my fish and photograph them eating for a story I am writing. I would write a song for my cat (animals love listening to their owners sing or play), play some games with a person I like, etc.

Conclusion

Many nations define their codes of conduct or accept core values proclaimed by other societies. They consider it imperative to fight with the sectarian, national, ethnic, linguistic, and racial attitudes and ensure that empirical facts, not ideology, drive the science. For example, integrity, impartiality, loyalty, accountability, and professionalism are needed for successful communication and collaboration.

Strategies concerning teaching, for example, inclusiveness, multicultural sharing, and cultural values, must be delivered clearly, according to existing environmental and social conditions. As the Millennials and Generation Z members show evidence of a short attention span and avoid reading long texts, the application of visual means of communicating any subject matter is preferable.

4 Thinking and creating

Cognitive, creative, metaphorical, and psychological capacities that support collaboration

Introduction

With the fresh approach offered in this book, retelling old tales and stories and reusing well-known characters is replaced with action-based stories designed for the respective audiences. High school students, those already building their careers, and their audiences and clients all want novel ways of getting in touch. New, not the retold stories with information hidden and interwoven into the exciting, even gripping plot and projects that embed knowledge, skills, and information serve as learning materials.

Art is essential to cooperative projects beyond the aesthetic experience we encounter in museums or galleries. Art and graphics exist in all disciplines, primarily visual communication media. Companies need to offer their audiences visuals (photos, video, comics, and more) and convincingly illustrated materials. Z. Nagin Cox acknowledged the visionary power of graphics when she revealed that NASA later produced objects created for *Star Wars* as real things, as their design met most of the standards. "No missions of space exploration would be possible without the technologies of computer graphics and the human/robotic interactions they enable" (Cox, 2016). A spacecraft TIE (the twin ion engine) fighter from *Star Wars*, made many decades ago in 1977, may serve as an example of how George Lucas introduced art that functions for producers after so many years (Microsoft Bing, 2023). Experiences of this kind support the conclusion that traits such as talent, individuality, and playfulness, when displayed by students, should be cherished as the best values that secure our future.

The themes discussed in this chapter and supported by learning projects include:

1 Artificial intelligence literacy based on watching changing daily progress
2 Visual, aural, and technological literacy
3 Cognitive thinking, information processing, and sensory, working, and long-term memory
4 Abstract thinking considering concepts beyond physical objects, such as freedom or love

DOI: 10.4324/9781032705347-4

5 Intuition: immediate understanding of concepts and situations without conscious reasoning
6 Visual imagination that helps to combine experiences, knowledge, ideas, and perception through the senses into one's own insightful representation
7 Creative thinking, concepts of flow, learners' creative development in problem-solving, problem-finding abilities, and digital creativity in technologies and information technology. Finding inspiration in nature for novelty without borrowing, copying, rephrasing, or mimicking somebody else's work.
8 Metaphorical thinking
9 Semiotics
10 Intelligence and individual intelligences
11 Spatial abilities

Artificial intelligence literacy

Somebody observed that the Millennials live on the internet, Generation Z members spend their time on phones, and Generation Alpha will prefer OpenAI systems and ChatGPT. People who suffered psychological and psychiatric problems during the COVID-19 pandemic because of a compulsory separation from others, now dream about starting their businesses and companies to work at home by themselves. AI techniques offer tools for photographic production, imaging, and writing. Users can quickly profit from applying AI writers, content writing tools, AI writing assistants, AI prompts, and text-generator services. The AI-based systems and applications advertise themselves as producers of plagiarism-free, unique, high-quality articles, essays, systems, and much more. Public support for these advantages and recommendations for AI-based apps meets controversial responses from people who see the societal-scale risks from AI and work on limiting its use and shaping new legislation. It is essential to remember that users are the authors of photographic or written works, and the AI machine is only a tool and an assistant. In contrast, users create projects with their sensitive, playful, and imaginative minds.

OpenAI

OpenAI is a research unit that derives its solutions from deep learning and natural language processing. AI-based machines learn from experience and their users and adapt to new tasks to perform some projects as well as or often even better than humans. However, we cannot expect intuition, sensitivity, or imagination from a machine. We have to train them to carry out particular tasks. Investors, big tech and start-up companies, driverless cars, and robots need AI technologies. OpenAI offers new jobs and careers for people from many disciplines and backgrounds and looks for audacious, thoughtful, unpretentious, collaborative, and growth-oriented candidates. Applications for AI for different fields of research and production grow daily. For example, the MIT Jameel Clinic's (*www.jclinic.mit.edu/*) faculty developed, with the assistance of AI, a new antibiotic named abaucin for treatment against *Acinetobacter baumannii*, one of the biggest threats to humans, which can combat drug-resistant infections (Trafton, 2023).

GPT

GPT (generative pretrained transformer) is a deep learning model originated by Google in 2017.

ChatGPT

In 2023, OpenAI publicly launched the ChatGPT app for iOS. A chatbot is a computer program that uses artificial intelligence (AI) and natural language processing (NLP) to understand customer questions and automate responses to them, simulating human conversation (*www.ibm.com/topics/chatbots*). The ChatGPT app syncs user conversations, supports voice input, and improves the model. A conversational AI can chat with users, answer follow-up questions, and challenge incorrect assumptions (*https://openai.com/blog/chatgpt-plus*). To use ChatGPT, one has to sign up and create an account in OpenAI. Users can type into the search bar, ask a question, or describe what they need: a poem, lyrics for a song, a video, information about a research paper, and more. They can give the ChatGPT prompts, possibly many detailed ones, and add the negative ones, telling what they do not want to happen in their work. Users should check that the answer does not contain incorrect, nonsensical, biased, or harmful content. They can add more prompts, and ChatGPT will understand or even remember what they wanted before this request. When searching for images using AI, some GPT models can learn the user's aesthetic taste and adapt their outcome.

GPT-4

To scale up deep learning and learn from human feedback, this multimodal model accepts prompts of images and texts as inputs and gives text outputs, such as natural language or code. It may produce human-level professional and academic performance. For example, it can pass a simulated bar exam with a score around the top 10% of human test takers, SAT, GRE, AP exams, and more. It can also accept text and images, including documents with text and photographs, diagrams, or screenshots (*https://openai.com/research/gpt-4*). GPT-4 is collaborative when it writes with users, edits and iterates the creative and technical writings, composes songs, writes screenplays, and even learns a user's writing style (*https://openai.com/product/gpt-4*). The versatility of

ChatGPT provides users with instant answers and information, guidance, inspiration, professional feedback, technical assistance, and learning opportunities (*https://openai.com/blog/introducing-the-chatgpt-app-for-ios*).

DALL·E 2

This AI system is a version of GPT-3 that can create realistic images and art from text descriptions (*https://openai.com/product/dall-e-2*, *https://openai. com/research/dall-e*). With GPT-3, language can instruct a neural network to generate text. With Image GPT, the neural network can generate images. DALL·E 2 can do inpainting by editing changes within a generated or uploaded image. Also, it can perform outpainting, where users may work on an image beyond its original form. By adding a description of changes, users can add visuals in the same style or develop a story in a new direction (*https://openai. com/blog/dall-e-introducing-outpainting*). However, users must be careful to avoid problems with copyright issues.

API

API – application programming interface – is a software interface that allows communication between computers and/or programs.

Cobots

Cobots are robots that collaborate with people. They are easy to program and install, have sensors that react to changes, and can learn new tasks at all skill levels and budgets. Cobots advance automation of production. They control machines' work, e.g., perform welding, offer more speed, and perform tedious and hazardous work so that people can do creative work.

EMBODIED INTELLIGENCE

Human cognition is not locked in a cerebral cortex but is also influenced by a body and experiences in the external world. Embodied cognition means that not only the mind guides the body, but the body also affects the mind (McNerney, 2011). This concept was studied in the 20th century by Martin Heidegger, John Dewey, George Lakoff, and many others. Lakoff and Johnson (1980/2003) wrote a book entitled *Metaphors We Live By* and posed that metaphors are concepts derived from nature and are represented physically in the brain. Another researcher wrote that mathematics is grounded in the body, embodying metaphorical thought (McNerney, 2011), giving tangible forms to ideas. With AI and the neural network framework, virtual robots have embodied artificial intelligence. They can receive visual and acoustic input and interact with other virtual robots and the virtual world. Then, these solutions are transferred to real-world robots.

AVATARS

Avatars are characters created as embodied agents in the real world as robots. They can learn and mimic human movements. Humanoids – humanized robots – have good kinesthetic abilities but not the ability to learn new knowledge and skills. The humanoid mind is considered an AI; its body is a robot. Embodied AI has many applications, including clinical therapeutic programs.

Using AI applications

Consider which system and app you need to carry out a project that would satisfy you. Consider options current technologies offer, such as the time and energy-saving shortcuts for your goals to help unleash your creative research-based work. With the support of AI, you may feel as if you had two brains – one artificial, which is informative, and the second one your own, creative and playful. Ask yourself who your audience is and what elements you need to successfully deliver your beautiful, explanatory, and well-designed solution. Ask questions and then type prompts to generate outcomes meeting your audience's emotions, needs, and expectations. Think about the format of your work and decide if you need help with your research, images, solving your questions and goals, or the time-based aids for your animation, video, VR, etc. Remember that when outcomes are provided to you, it will be up to you to edit, incorporate, or cut according to your creative thinking. You are the author of the project. The elements are offered to you, and you decide which ones you will accept, reject, rework, use partially, ignore, or eliminate. If the outcome doesn't satisfy you, change your prompts. To do it effectively, an artist or designer should imagine how the machine reads and responds to human thinking.

With what message and moral would you like to leave your audience? You are the one making decisions. Remember that many AI companies offer services, so their results may differ. Also, there are lots of copyright-free resources on the internet offered to you at no cost—for example, textures.com, Pexels, and more. One can contribute to or take from. Read each company's policy about rights. It is always nice to give proper credit to acknowledge somebody else's work or show appreciation for a company that hosts such a collaboration. Finally, think about the role of a text, type in your work, and provide an artist's statement.

Visual, aural, and technological literacy

Visual, aural, and technological literacy are helpful for successfully designing integrative projects for knowledge transfer through visual learning. Learners become familiar with traditional media and know a little about new media arts created through the web, social networking, smartphones, tablets, and some types of watches.

Visual literacy

With visual literacy, one can read, write, and create visual images. It has been described as the ability to create, evaluate, and apply conceptual visual representations and see how they depend on the balance between them. While visual learning was already practiced in ancient and medieval times, the term 'visual literacy' was first introduced in 1969, when the International Visual Literacy Association was founded (IVLA, 2023). Visual literacy allows us to see and interpret images and use visuals to convey meaning. Researchers study the visual part of the brain, its plasticity, the organization of neurons, and how the nervous system receives, processes, and visually transforms information. We can translate scientific concepts into mental imagery and visual thinking. The import of visual literacy grows because of the increasing presence of communication media.

The ability to visually express scientific and technical concepts depends on our balance between visual and technological literacy. Instruction in science and technology that uses the visual way of learning improves learners' visual literacy. While examining mathematical equations, Albert Einstein could visualize the physical reality behind them, while many people could only see equations as a series of logical codes. For him, math was nature's playbook; he saw gravity as geometry (Isaacson, 2021). Today, it is even more worthwhile to remember that product design, communication media, marketing, and advertising strongly depend on visual perception, associations, metaphors, connotations, color relationships, and other aspects of a designer's thinking.

Individual people may have different responses to both natural environments and artworks. When approaching Niagara Falls, someone may feel an intense aesthetic experience, while another person may complain about the cold and wet wind. In a museum, the first person may feel elated and respond emotionally to artwork, while the second may feel tired, hungry, or need to check messages on their phone or watch. The artist and the viewers may understand the artwork differently. As viewers, we often do not know the artist's message, as in the case of prehistoric cave drawings, which most of us perceive as beautiful. There are many theories about art's meaning and the reasons for art creation:

1 A warning for others, giving ethical, moral, or other messages
2 A way of processing and reporting information

3 Expression of the artist's attitudes, emotions, philosophy
4 Outlet for one's creativity
5 Aesthetic experience or decoration
6 A way to commemorate and immortalize oneself
7 Relieve tension and anxiety
8 Fulfillment of unconscious thoughts
9 A vehicle for truth or knowledge
10 Art for art's sake
11 Representation or a record of reality
12 A way of showing the concept of unity or variety
13 A hedonistic fulfillment
14 Many new theories about art meaning are coming frequently

Aural literacy

This ability characterizes people who prefer hearing something to examining it visually. The aural preference may determine the learning style of the aural person. According to Drinko (2021), aural learners prefer to hear things aloud, gravitate towards audiobooks, close their eyes to understand something better and focus on auditory inputs, and talk to themselves or mouth the words as they read. They are good at learning people's names, do not like competing noises, and may not pay attention to visual information such as charts and graphs.

Technological literacy

In many cases, small children can use software that teaches them about science, mathematics, programming, and visual storytelling. The role of networked technology in artwork creation is growing. The same artwork can be projected on a huge screen on a wall or seen on one's watch, so the size of a work, often deemed not so important, may significantly change its meaning. One artist said, "I create postage stamp–size artworks because I have a tiny studio." Art critics are judging the aesthetics of digital culture on websites, web browser applications for accessing websites, software applications, and electronic art performances. Viewers discuss the new art forms while participating in polls and voting about a particular artwork.

We can visualize our thinking and think in pictures as well as verbalize it. A linear, sequential fashion is typical of talking. With images, linear information is translated into spatial metaphors. Graphics help us reason about presenting data, communicating, documenting, and protecting knowledge. Infographics such as lines, shapes, pie charts, bar graphs, or diagrams represent data, information, or knowledge. Isotypes (International System of Typographic Picture Education) show social, technological, biological, and historical items as pictures. They use standardized and abstracted pictorial symbols and identical figures in serial repetition.

Possibly, with visualization techniques, we can best learn, teach, or share data, information, and knowledge. Visualizations allow us to avoid dry texts that are difficult to absorb. We apply visualization when we design a presentation of abstract data in a visual, often interactive way, including via VR. For this purpose, we use well-known objects connected through well-defined relations. For example, popular presentations of atoms, molecules such as DNA, cells in living animals or plants, and of biological or chemical processes going there are represented in simple drawings that visualize complicated structures and processes.

Basic data literacy means the ability to analyze, interpret, and also question data, often big data (Harvard Business School Online, 2019). Data-literate professionals can handle, analyze, and interpret data, so they can make decisions.

DIMENSIONS

We can see and process objects or concepts in as many dimensions as we need. Dimension can be defined as the number of coordinates needed to specify a point on the object. Coordinates are numbers that indicate the position of a point, line, or plane. A line is one-dimensional (1D); we can measure points on a line by the distance of its points from zero along that dimension. With the points, we can make a straight line, a circle, a helix, or a curved outline. A rectangle is two-dimensional (2D), and a cube is tridimensional (3D). We can measure spacetime in four dimensions (4D), which are relative to our motion as observers rather than define it spatially and temporally. Tony Robbin (2006) made computer visualizations of four-dimensional geometry, hyperspace, and models of the fourth dimension and examined how they can be applied in art and physics.

Project 1 and Project 2 relate to visual, aural, and technological literacy.

Project 1. Two forms of creating texts and illustrations: an illustrated writing or a pictorial work with a description

This project proposes exploring relations between texts we are writing and images made to upgrade the text and give it better quality. The project comprises two parts:

a   First, it invites you to either work as an author-illustrator by creating an illustrated writing or begin with your pictorial work and then add descriptions.
b   The second task in this project comprises writing a critique of somebody's work.

Background information about authors-illustrators

Many writers apply their talents and creativity to illustrate their writings. Also, calligraphers and hand-lettering artists make words come alive. Chinese calligraphy was admired as a fine art long before painting and is still performed as that. During the Song dynasty (960–1279), calligraphy gained the high ranks of the fine arts, not only the status of a mere craft. A similar standing of calligraphy can be seen in the Arabic culture, both old and current.

Below are listed some writers who used to create pictures, drawings, or sketches along with their stories.

> Lewis Carrol (1832–1898), e.g., *Alice's Adventures in Wonderland* and *Through the Looking Glass*
> Kurt Vonnegut (1922–2007), rough felt-tip pen illustrations and doodles added to his prose
> Antoine de Saint-Exupéry (1900–44), novels and *The Little Prince*
> Henry Miller (1891–1980) created an estimated 2,000 watercolors
> A. A. Milne (1882–1956), *Winnie the Pooh* books and poems
> Dr. Seuss (1904–91), writer and illustrator
> Jim Henson (1936–90), a puppeteer who created the Muppets and much more
> Maurice Sendak (1928–2012), writer and illustrator
> Erik Carle (1929–2021), author, designer, and illustrator
> Chris Van Allsburg (born 1949), illustrator and writer
> Hayao Miyazaki (born 1941), animator, filmmaker, author, and manga artist
> Walt Disney (1901–1966) animator and film producer who wrote his stories
> William Steig (1907–2003), a cartoonist and children's books author, and illustrator
> Norton Juster (1929–2021) an architect, writer, and children's books author and illustrator; for example, he wrote *The Dot and the Line*
> And a great number of other writers.

Background information about a pictorial work with descriptions

Many painters and other visual artists also wrote essays, novels, plays, poetry, and screenplays. For example,

> William Blake (1757–1827), a painter and printmaker, was also a poet
> Pablo Picasso (1881–1973) wrote over 300 poems and two plays
> Salvador Dali (1904–1989) wrote and designed an animated short film for Walt Disney titled *Destino* and *Impressions of Upper Mongolia* about finding hallucinogenic mushrooms.
> Andy Warhol (*c*.1928–1987), a painter, graphic artist, and filmmaker, e.g., *Batman Dracula*
> Ben Shahn (1898–1969) published his lectures, titled *The Shape of Content*

Julian Schnabel (born 1951), a neo-expressionist painter and filmmaker, wrote and directed a film about the artist Jean-Michel Basquiat (1960–1988), a neo-expressionist artist

Crockett Johnson (1906–1975), a cartoonist and children's book author and illustrator, created over 100 paintings relating to mathematics and mathematical physics.

In many cases, the image created by an artist is connected to the music this artist (or someone else) composed. Many combinations of links were created between pictures, music, and texts (sometimes poetry). For example, Bob Ehle improvised his musical compositions to computer art graphics created by the author of this book.

Guides for working on the project

a   Making one's illustrated writing or a pictorial work with descriptions

This project aims to make visual thinking easier, enrich communication through the arts, and help us to recognize the values of visuals that often come in the form of the arts. When you finish the initial analysis, work on your project. This project is open; decide whether to begin with the visual or the written part. You may prefer to start with creating visual storytelling with drawings, animation, or a video and then write a short story or a verse. You may also prefer to work the other way around, starting with writing a story in prose or as a verse, all the time creating illustrations that support passages you have written. For instance, you may first describe an anxiety-ridden adventure in a stuck elevator, telling about the thoughts, exclamations, and actions caused by this trouble. Then, visualize this event by sketching yourself and other elevator users closed in a tight space for an unknown time period or making a short animation about this situation.

b   Writing a professional critique

Pick an artwork, music composition, or verse that is important to you and that you would like to learn more about. Go online to find information about the artist/author you selected, writing proper references about sources of information you found. Then, write a critique of the artwork created by this person. Talking about art and making an art critique may enhance higher-order thinking skills and stimulate creativity, visual imagination, spatial abilities, and visual literacy. Because art facilitates good communication, it may support sharing with classmates both your own projects and reactions to the works of others.

Describe your motifs and reasons for choosing the theme, and compare and contrast the strength and values of both the visual and verbal parts of someone's works. Our judging of artwork may depend on external factors that can bias our reaction to artwork and influence our aesthetic judgments: information about who the artist is, whether this artist is universally known, and whether the work is shown in a prestigious art gallery. The same artwork may evoke different, unique, and individual messages in viewers. We are taught to recognize 'good' (beautiful) and 'bad' (ugly) art and discriminate between the 'good' and the 'bad.' However, what is meaningless for someone can look beautiful or meaningful for another person. No two people share the very same opinion because they have different life experiences. Also, the aesthetic experience may be enhanced by the viewer's familiarity with the artwork if they have seen its reproductions online, in art books, slides, videos, and movies.

Critical analysis is considered a thinking strategy that involves description, analysis, interpretation, and judgment. While talking or writing about art, ask yourself four kinds of questions to evaluate what you see and experience:

The descriptive questions ask what we see.
The analytical questions ask what the interrelationships of the parts of the artwork are.
The interpretative questions ask what the meaning is to you; and
The evaluative questions ask whether it is good or bad art and whether you like or dislike it.

Write a critique of an artwork following the guides. This skill should help you at social gatherings, possibly even at job interviews, as many employers see a need to include visual communication abilities.

* Make a description of the artwork under critique; the artist's name, the title of the artwork, its size, medium, and date of production. First, collect the facts only (not your opinions) relating to depiction, composition, placement of objects, technique used, etc.
* Analyze and describe the facts you have collected by looking for the proportion of forms, perspective used in the work, use of light and shade effects, resemblance to the subject depicted or its deformation, and relations between forms and the background.
* Try to make a statement that would unify the traits of the work. Make a critical interpretation of the observations you have made. Do the work's particular traits combine to fit together and make sense? Make a judgment on the work in terms of the formal beauty of the artwork, expressive values in communication with the viewer, and instrumental effectiveness for a purpose. Justify your opinion on the work under critique.
* Explain why you like this work. Why did you enjoy looking at this work, and if you prefer this one to the others, explain why; if not, justify your

opinion. The reasons could involve connotations, associations, memories, analysis of color combination, relationships of some parts of the work, its mood, or a type of message it sends to you.

Take care to fulfill the requirements of writing a critique. Type your critique as clear, concise writing. You should write at least 80% of your critique in your own words. Critique should be based on your research. Your writing should give the reader strong evidence that you have conducted a scholarly investigation of the artwork. Make sure that all four of the descriptive questions are answered.

No strict rules or definitions would tell you whether your or somebody else's work is good. There is no definition of art because we cannot fulfill the "all but only" condition: We cannot say *all* artwork is beautiful because we all saw good but ugly artwork. We cannot say *only* artwork is beautiful because Nature has much beauty. You have probably experienced one ugly but very good work of art, and you admired it nevertheless. It becomes even harder to state if art is good or bad because it is a personal judgment. My definition of good art is: place your artwork on a wall or in the center of your room for a month. If you can live with it and feel comfortable, you may be confident when you decide to show your work; if not, work on it.

SUGGESTIONS FOR INSTRUCTORS

It may be useful to provide a guide for the learners:

1   Saved or exported to Microsoft Word or a PDF
2   Critique should be based on your own research
3   80% written in your own words
4   All descriptive questions are answered.

Project 2. Writing a visually appealing research-based poster

Background information about research-based posters

The academic research poster is a format seen at many conferences. It should not be confused with a poster as an art form. There are many poster sessions at conferences and meetings where authors of posters may provide visual information about their work. The research-based poster design is typically developed to convey some message or a fact in a graphical way. It requires a specific set of skills to make it bold, easy to understand, and attractive for the eye of a beholder yet to allow the audience to compare and contrast data presented visually. Many free PowerPoint templates for research poster presentations can be edited for personal purposes. A poster may also present visual information

Table 4.1 Rubrics for evaluating a critique

Rubric for The Art Critique	Exceptional 20–18 points	Conscientious 17–14 Points	Mediocre 13–10 Points	Weak 9–0 Points
1. Quality of Research and Content	Excellent evidence of authentic research and content utilizing quality resources, including your own highly developed ideas.	Good evidence of research, and good content including some of your own ideas.	Limited evidence of research, limited content. Basically, a regurgitation of information.	No evidence of authentic research.
2. Attention to detail and explanation of the four questions	Excellent attention to detail and exploration of the four questions, written with interest.	Sufficient attention to detail, with a conscientious exploration of the four questions.	Limited details, only basic information revealed.	Not nearly enough detail. Does not fulfill requirements.
3. Proper format: double spaced. File named according to the specification (Your_Name_Critique)	Proper format and required image. The four questions made bold and numbered. Your research-based answers provided. References follow.	Good use of format but missed one of the requirements.	Too many requirements for proper format missed. No references.	You missed the majority of research requirements.
4. Thesis: Clearly states a thesis, which is supported in the Critique. Has a conclusion.	Paper exceptionally describes a thesis and provides strong evidence to defend the thesis.	Your Critique conscientiously addresses and supports the thesis.	Your Critique makes a faint attempt at addressing the thesis.	No effort to form a thesis or defend the thesis.
5. Quality of writing, including proper punctuation, sentence structure, proofreading, spellcheck, etc.	Excellent writing quality. Well-written, innovative, and interesting to read.	Good writing quality, with attention to detail and interesting points.	Acceptable writing. Your paper would benefit from using writing resources.	Poor writing. Not written to college standards. Please use writing resources in future.

about your current and future professional profile. This project aims to determine what you have learned about a selected area of knowledge and indicate your goals, strengths, and career readiness. This project should be prepared in a formal style and should contain citations and references.

Guides for working on the project

You may use data graphics and other visual materials you made before following your interests and tasks. Choose a leading theme for your presentation and collect old and new materials supporting your illustrated writing. After choosing a theme, consider the important message you want to tell your audience and design a layout to emphasize this part of your presentation. Take as many photographs and design graphics, such as graphs and tables, as possible to make your presentation visual and avoid copyright problems. Compose a descriptive title and write it in a font size visible from far away. Prepare your images and texts large enough to be read from 5 to 10 feet away. Choose an easily readable font and color. Think about a background that would be compatible with your poster content. Make your images contrasted, well related to the text, and eye-catching. Write your text concisely with sentences that are short and not too detailed.

Make your research poster attractive by adding color. Choose a beautiful color scheme and a well-balanced layout. Think about color coordination to make your message more clear. For example, you may apply color coding to arrange the items you are discussing there, but remember to provide a legend explaining your use of colors.

Apply colorful headings, bullets, and graphics that would break up and organize your text. Think about your audience; predict their questions and include the answers. Carefully check for typos and mistakes so the text is error free. Remember to insert your contact information to enable communication with you.

Cognitive thinking skills

To make decisions, one has to search and choose needed information to enrich one's memory store and retrieve valuable data from there. In the cognitive process, learners use knowledge, meaning, and understanding to shape their way of thinking. Thus, cognitive thinking is essential for creating new projects, studying, performing research, testing and interpreting new information, and creating something new. For cognitivists, the brain acts as a computer to process information.

Three main domains of learning are the cognitive (thinking), the affective (social/emotional/feeling), and the psychomotor (physical/kinesthetic) domain. People learn using these cognitive, psychomotor, and affective domains. They activate their preferred learning styles that depend on their visual, auditory, kinesthetic, or reading/writing preferences. Projects offered in this chapter are designed to address the three main learning domains. Hopefully, they may serve instructors in designing new holistic educational tasks.

According to the framework reworked by Benjamin Bloom and his collaborator, David Krathwohl, Taxonomies in the Cognitive Domain consisted of eight major categories: knowledge, comprehension, application, analysis, synthesis, evaluation, induction, and deduction capabilities. According to Leslie Owen Wilson (2020), the taxonomy revised by Lorin Anderson, David Krathwohl, and Benjamin Bloom offered in 2001 (Anderson et al., 2000) comprises:

1 Remembering: drawing knowledge from memory, retrieving previous information;
2 Understanding: building meaning from written or graphic messages and/ or activities such as interpreting, exemplifying, classifying, summarizing, inferring, comparing, or explaining;
3 Applying: using material learned from models, presentations, simulations, or interviews;
4 Analyzing: dividing concepts and materials into segments, then organizing and examining how they relate to the whole structure by creating spreadsheets, surveys, charts, diagrams, or graphic representations;
5 Evaluating: using standards and criteria to check and judge, to produce critiques, recommendations, and reports;
6 Creating: making a logical or functional unit reorganized and synthesized in a new way through generating, planning, or producing. Creating a new form or product is the most difficult mental effort.

Many disciplines study the mind, such as psychology, linguistics, computer science, philosophy, anthropology, and cognitive neuroscience, along with general medicine, criminal justice, sports, and health sciences. Scientists design computational models of cognition. Cognitive science studies the mind and its processes, examining the nature, the tasks, and the functions of mental actions. Cognitive thinking informs us what we know and relates to problem-solving and making new concepts. People use cognitive thinking when they analyze information, organize it into a memory store, and then retrieve information from memory to make associations, connotations, and then educated guesses. They use it, not necessarily consciously, for decision-making, gathering data and information in experimental research, testing hypotheses, interpreting results, and providing scientific evidence for new theories. We perform cognitive activity when we make calculations, compute, recognize symbols and signs, read spatial maps, and write poetry, but we also bring past things to mind or recall old names and events.

New information, including visuals, goes through three phases in the human brain: sensory memory, working memory (temporarily holding new information), and long-term memory. First, short-term memory gathers some information for a limited time to process and store it in long-term memory. Some psychologists call the attention window the breadth of attention, which means the size of a display we can perceive and discern. To examine how long-term memory works for you, draw a mental map of the place you lived as a child and check out how much you remember.

Activity: memory, visual memory

Ask three unrelated individuals on the street for directions to the same place. See if the three descriptions match. Would someone ask you if you have a smartphone with a GPS mapping software to find a way to this place and use the best version of the way?

Humans are a part of the animal world, in which many species present extensive consciousness, emotions, feelings, empathy, playfulness, and a sense of humor. These faculties were confirmed in animals by studies on the behavior and personalities of elephants, dolphins, whales, wolves, dogs, big cats, apes, and other species (Safina, 2015; Worrall, 2015). Our mind depends on our brain's functioning, which uses specialized cells called neurons. Neurons, mostly located in the central nervous system, send signals to other neurons, muscles, different cells, and the outside world. Electrical and chemical signals going outside and toward the brain are transmitted and transformed through synapses, the specialized areas at the ends of the neuronal extensions called axons. The brain has about 86 billion individual nerve cells and trillions of synaptic contacts. Large neural networks act as neural circuits consisting of smaller circuits. Patterns of activity in the large networks respond to signals from the body, the external world, and the patterns of other circuits in the brain.

As a result, our conscious and curious minds can learn different things from different people. They may work within a pragmatic (related to where we live), perceptual (showing what we experience), existential (introducing social and cultural issues), cognitive (based on thinking), and logical (abstract) space. Mental and cognitive operations are going by the activation of neuronal networks. They are registered in many ways, for example, by functional magnetic resonance imaging (fMRI), which measures changes in the cerebral blood flow resulting from activating a region under observation.

Learning styles may depend on individual preferences, and they activate different parts of the brain. They depend greatly on different types of intelligence, described by Howard Gardner (1993/2011, 1993/2006). Gardner discerned visual/spatial intelligence, verbal/linguistic, logical/mathematical, bodily/kinesthetic, musical/rhythmic, interpersonal, intrapersonal, naturalistic, and existential intelligence. The preferred learning style may be primarily visual (and spatial), aural (auditory and musical), verbal (linguistic, in speech or writing), physical (kinesthetic), logical (and mathematical), social (interpersonal), or solitary (intrapersonal). Howard Gardner believes everyone possesses more than one strong intelligence (Gardner, 2000).

Project 3, Project 4, and Project 5 relate to the cognitive activities section and invite the readers to make visual stories related to nature or science.

Project 3. A biology-based story

In this biology-based project, tell about life in a colony of animals or dramatize a butterfly's life cycle. Create and engage in action several characters, one for a colorful egg and other ones for the following stages of the insect's life: a hungry caterpillar (larva) growing and shedding its skin several times, a chrysalis rapidly changing inside a pupa, and a dramatically emerging adult butterfly which has to dry its wings to fly away and try to find its mate. Introduce drama by including threats to butterflies: snakes, frogs, birds (such as black-backed orioles and black-headed grosbeaks), spiders, wasps, dragonflies, lizards, ants, stink bugs, mantids, and even ladybugs, all of which to eat butterflies, especially monarchs. This project does not aim at presenting the exact features of biological species. Each character may look imaginary and tell about its life without showing its actual form or structure.

Project 4. A physics-related project about four states of matter

In this project, display how changes in the state of matter (solid, liquid, gas, and plasma) depend on factors such as temperature and pressure. You may present states of matter as characters, created and then animated. A solid character would have a definite shape and volume, while a liquid one, with a definite volume, takes the form of its container (possibly with comic effects). A gas character lacks a defined shape or volume (with dynamic, colorful effects of moving particles), and a plasma character shows its electric charge (with dynamic effects using colors and action).

Project 5. Playing cards as a metaphor for a company's structure

In a visualization showing a business-oriented line of work, you can create characters as a set of playing cards. This project can serve as a metaphor for the real-life relationships presented routinely as a flow chart and showing the roles of people involved in a complex system of a company structure. Draw eighteen characters for face cards in a set of playing cards:

1 Ace, king, queen, jack, and two jokers.
2 Make them for four suits: diamonds (♦), clubs (♣), hearts (♥), and spades (♠); thus, you will have 18 characters.
3 Sketch those characters with a hierarchy: from the most important to the least power your character possesses.
4 Face cards: a king, queen, or jack are characters that determine each suit.

Draw the face cards as characters designed according to your interests. For instance, playing cards may present themes such as particular sports, hobbies, cars, gardening, animals, human works, music, cultural issues, geometry, etc. Also, design the back of the deck of cards, and define typography as a vehicle for successful communication about the theme you chose when you drew the cards. Then, discuss your characters, the approach you chose to their creations, and the roles they play in storytelling about the fates of the cards.

As a continuation, create an organizational chart showing a company's structure, a fictional or a real business of your choice. Use visual information (obeying the copyright restrictions) about the organization of this flow chart, and use metaphors for a company structure: for example, president, CEO, and CIO may be the ace, the king, and the queen. Present other employees as jack, joker, Card No. 5, etc.). Short stories about each character will tell about its placement in this imaginary thematic society.

Abstract thinking skills

Abstract thinking allows us to consider concepts beyond physical objects and experiences we physically encounter. Abstract concepts like freedom and love are real but not directly tied to the physical world. The development of abstract thought begins in the kindergarten years. In their imaginative, dramatic play, children use objects to pretend they are other things and use them symbolically with a high level of abstraction. For example, they may pretend they are driving a car when they sit in a box or use a wooden block as a cell phone. They learn how to apply symbols: they use a mental image of an object to substitute one object for another or mime an object or action. Then, children learn that letters and numbers represent something. The ability to abstract thinking makes learning more accessible in all curriculum areas. Applying abstract thinking in the process of learning makes concrete themes for studies more interesting.

Drawing can make abstract thinking and understanding an idea easier because images, symbols, icons, metaphors, analogies, and words lie along a continuum, from concrete and representational to highly abstract. Abstract drawing may simplify an object depicted, by, for example, converting images into geometric shapes or forms (Figure 4.1). A photograph of a desk lamp or a car is a concrete representation, while an electric circuit or a blueprint is abstract. A Slippery-when-wet road sign usually depicts a driver, but in some countries, it has not, even before self-driving cars. A sign about the availability of a phone communication usually shows an iconic handset to be held up to speak into it and listen to instead of a current smartphone.

We may find many writings done abstractly, conveying thought without describing objects or events, for example, the short verse "A Win" below.

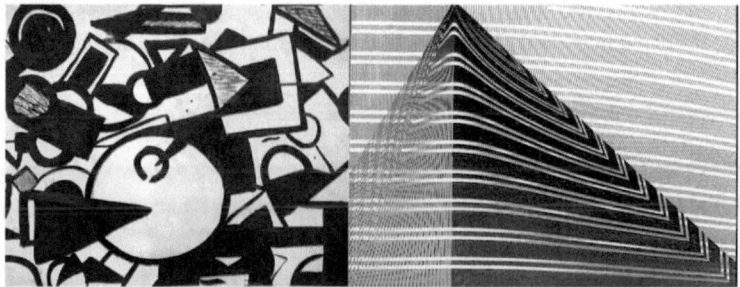

Figure 4.1 Two abstract images, one drawn with ink on paper and the second programmed on a computer

A Win

My speller corrected my ChatGPT.
How decadent!
The End.

Activity: creating an abstract art

Collect six random pieces of paper. You may choose random shapes (with rounded edges) or prefer geometric shapes of paper, for example, rectangles: big or small, straight or tilted, flowing, folded, or colored rectangles. Set them up as a composition you would enjoy looking at. Decide if you want to give a name to your composition to direct viewers' imagination or consider it pure abstract art.

Due to abstraction, we enhance features that best identify an object and suppress those not necessary to comprehend it. To give our drawing meaning, we avoid features that are not crucial for representation. Highly abstracted, not representational drawings become symbols. Symbolic drawings may convey messages about universal forces.

Abstract, nonrepresentational art often uses symbols and symbolic drawings. Below are artworks that progressively abstract the essence of the picture theme.

- Photography shows everything we see, unimportant and important things, as in Anselm Adams's photographs.
- A portrait painted by an artist enhances characteristics specific for a sitter.
- A caricature deliberately exaggerates the subject's distinctive features or peculiarities. It often produces a comic or grotesque effect.

- Artists who developed conceptual, nonobjective, abstract art, abstract expressionism, or minimalist art were abstracting features and issues they considered important to the point that objects were becoming beyond shape recognition.

Activity: abstract/representational drawing and the other way around

1   Geometric/organic drawing

Draw a geometric shape and then make an animal out of it. For example, a circle or an oval may transform itself into a ladybug; a hexagon – a plane figure with six straight sides and angles – may evolve into a crab; a rhomboidal shape with four equal angles and four equal sides may resemble a frog; and a curved, thick line may change into a snake. Find other geometric shapes that will resemble animals when you imagine them.

2   Realistic/nonrepresentational drawing

Focus on an object that is in front of you. Convert it into nonrepresentational shapes that will combine into the form of this object. Name these shapes. Try first with those having abstract, irregular features, then focus on the geometric shapes. Define the difference between the notions of geometric and abstract shapes.

3   Find an object and change its meaning.

a   **Object.** Select any object taken out of your pocket. Make a sketch by emphasizing its most characteristic features. Save your file: YourName_Object

b   **New object.** Change the name of your file to YourName_New_Object. Look at your initial sketch. What does it remind you of? What type of connotations or associations can you make? Add new meanings to an object so that you will convey different messages through it. Redraw or transform this object to become something else and gain a new meaning.

For example, if you have a Kleenex in your pocket, you may change it into a sail by reshaping its surface and adding a hull. If you carry a key, you may shape it into a folding knife.

c   **Objects as characters.** Then, change the meaning of this object again, making it a character for a storytelling project. Make simple sketches showing your old and new objects as two characters. Imagine animation that you could make using your sketches of characters. For creating your story, use sketches as a starting point. For two sets of your drawings: sketches of the object and sketches with its changed meaning; assign various feelings and expressions that will serve later in the story development. Your characters will express moods and convey messages through body language and facial expressions. For example, they may convey joy, show anxiety or fear; they may be warning or inviting, friendly or aggressive. Now, you are ready to construct a storyline.

In the times of networking, abstract concepts presented as art, often using computing, can be seen in many kinds of artistic competitions, for example:

- Joint Mathematical Meetings *(https://jointmathematicsmeetings.org/jmm)*
- Bridges: the Mathematical Connections in Art, Music, and Science (*www.bridgesmathart.org*) and Mathematical Imagery at the American Mathematical Society (*www.ams.org/home/page*)
- NanoArt International Festivals, Competitions, and NanoArt21 Exhibitions (nanoart21.org)
- Fractal Art Contests (*www.fractalartcontests.com/*)
- Nikon's Small World Photomicrography Competitions (*www.nikonsmallworld.com/*)
- D-ART Art Gallery for the International Conference on Information Visualization, South Bank University, London, and GraphicsLink, GB

Project 6. Love as a force and power

Background information

Picturing abstract ideas gives us an insight into concepts we cannot see, but we know them well and often use them as metaphors. Love is one such abstract concept. Think of love as a power, force, and building material. There have been many threats to humanity, from natural disasters such as earthquakes, volcano eruptions, fires, and floods to disease outbreaks, wars, crime, or violence. With love, we understand and protect people and their values. Love gives us strength to perform deeds and motivates us to develop knowledge, produce drugs and vaccines, and initiate social changes. It is the enigmatic force that pushed the members of the United Nations to adopt 17 sustainable development goals to preserve the natural world (https://sdgs.un.org/goals, 2023). In Robert J. Sternberg's (1986) triangular theory of love, the three components of love, in the context of interpersonal relationships, are intimacy, passion, and commitment.

There are many conflicting issues related to love. The doers need to establish order, create revolutions, and change values. The invention of gunpowder caused many deaths. The Nobel Prize was established to balance the dreadful outcomes of regrettable inventions. We do not need wild animals for food, and we preserve endangered species. Some issues have been going around in circles throughout history, like good and evil fighting in a circular motion. We wonder how AI will affect us and its inventors' debate on related rules and restrictions.

Guides for creating the project

To visualize an abstract idea of love, think about how love solved problems, cured, healed, and helped people to survive. Find a convincing example of how, over centuries and across the globe, love made people stronger, nurtured them, and helped resolve conflicts. Avoid the heart symbol or any specific symbol related to rituals or an individual experience. There are too many religions across time to show them all, so try to avoid specific references that would narrow your story down to one culture or religion.

Convey visually the essence of the message, delivering this project through the arts, graphic design, visual storytelling, or video/animation. A message about the power of love may be clearly identified, or it may be open to more than one interpretation. Images with iconic properties and generally accepted symbols are used in knowledge visualization. Cognitive science specialists consider metaphors inspired by nature as instinctively understandable concepts.

Craft your composition as a general idea rather than a personal story or a tale about "how a boy met a girl, and they lived happily ever after." Work on designing an icon for love's power. Icons and iconic objects are familiar to all viewers, for example, math symbols such as more $>$ and less $<$ or not equal $\neq$. Many icons stand for abstract ideas. For example, an old-style electric bulb shining above a head often stands for a bright idea. However, many ideas and concepts do not yet have their iconic images, for example, intelligence, not to mention artificial intelligence, machine learning, and deep learning, which applies neural networks. Many icons have a long life. We do not use old-style phones with handsets, but we can see this icon on many interstate highways. In the same way, a picture of an anthropomorphic humanoid robot still serves as an icon for robots designed for specific tasks or bots (virtual software agents) that do not resemble humans.

Activity: Sketching icons

Create an icon that would signal some specific function, situation, or event that is typical of you and your everyday tasks. Then, create a series of icons symbolizing, for example, your typical daily activities from morning till night.

Intuition

The New Oxford American Dictionary informs us that intuition is "the ability to understand something immediately, without the need for conscious reasoning." Intuition, a superior capability of experts, is a uniquely human skill. It means one can answer without being able to explain how one knows that answer, which creates a big liability in terms of artificial intelligence techniques. Our ability to provide rapid, accurate answers without a need for rational thinking depends on automatic, quick information processing occurring not only in the cerebral cortex but in the caudate nucleus, a small area in the brain's basal ganglia (Wan et al., 2012), a nerve center of learning and automatic behaviors.

Intuition, considered by many the highest form of intelligence (Kasanoff, 2017), develops after long-term training that results in knowledge (Koch, 2015). Intuitive answers not only depend on our brain but also are embodied by using signals coming from our body and a history of our experiences in the external world.

It seems intuition, as a form of cognition, acts as the nonverbal thinking not necessarily involving conscious awareness. However, characteristics of the consciousness and our awareness of self and the world are subjects of discussion. Researchers examining the physical, neural, cognitive, functional, representational, and higher-order aspects of consciousness discuss whether our consciousness relies on the brain's local or general activity (Van Gulch, 2014). Because nonverbal thinking does not depend on language, it may occur in a nonlinear way without making step-by-step deductions, and it may happen faster than verbal thinking. Without the ability to communicate verbally, animals react much faster than people to external events. In many cases, nonverbal thinking uses pictures to visualize objects of one's thoughts and compare them, often without conscious attention to visual and sensory patterns already existing in one's memory. The ability to gain intuitive knowledge about a person or thing is called insight.

Our nonverbal communication uses images (signs, posters, or pictures), sounds (speech or music), time-based movements like gestures (the everyday expression to enhance our statements, and also mudra gestures in Hinduism and Buddhism), or artistic forms such as actors' gestures, body language, and motions, among other possibilities. Cherry and Susman (2023) list several forms of nonverbal communication, such as facial expressions, gestures, paralinguistics (such as loudness or tone of voice), body language, proxemics or personal space, eye gaze, haptics (touch), appearance, and using artifacts (objects and images). We are only sometimes consciously aware of these subtle messages.

Our embodied cognition allows us to use visual metaphors of concepts that are derived from nature and represented in the brain regions (Lakoff & Johnson, 1980/2003). Visual metaphors contribute to our intuitive decisions that are unrelated to language when it comes to planning subsequent actions, responding to social situations, reacting with empathy to other people troubles, making moral evaluations, and acting in accordance with our self-image. Many trust their intuition when detecting deception, lies, hypocrisy, or understanding other people's true feelings rather than looking for analytic reasons for their judgment. Thus, intuition helps us to make sense of reality and react spontaneously without using words.

These remarks may refer to our sight and our haptic skills relating to the sense of touch. There is a tale about a famous eye surgeon, Dr. Vladimir Petrovich Filatov (1875–1956), who used to ask his patients how deep they wanted him to cut a stack of thin tissues: He asked 20, 30, or 50 layers? The doctor then made incisions in a stack of tissues exactly at that expected depth.

Ancient and medieval physicians had to develop the haptic ability and skill to correctly diagnose an illness and judge its cause, duration, and prognosis by examining up to 30 kinds of arterial pulse beats' rate and shape. They felt the volume, strength, weakness, regularity, or interruption of the superficial, deep, slow, quick, saw-edged, undulating, and worm-like pulse (Ghasemzadeh & Safari, 2011; Zoupan, 1760). Currently, physicians often do not even touch patients who come to a clinic or a hospital because they believe the data gathered by the high-technology instruments and devices are enough sources of information. They receive a numerical output but not an insight from their sensory perception.

Activity: Sketching actions

Imagine five adventures with dangerous situations that may happen to you. Each time you visualize the danger immediately and spontaneously, make a quick sketch of your reaction to the scary situation. What would you do to save yourself? After that, write your rational reaction to this imaginary danger: What would you do, why, and how would you act in the most clever, intelligent way? Will your intuitive and reasoned reactions to the same challenge be consistent and similar? You may then ask the AI GPT-4 app for advice in one of these situations or ask it to sketch it for you. You will receive an answer based on the data analysis of similar cases, completely devoid of any unconscious insight into danger recognition, as the AI machines do not have consciousness yet.

Visual imagination

With imagination, our mind can combine experiences, knowledge, ideas, and concepts into one's own insightful representation. Imagination may relate to our perception through the senses: sight, hearing, taste, smell, touch, balance; kinesthesia, which gives us a sense of motion and acceleration; proprioception, which allows sensing the relative position of parts of the body, feeling direction, temperature, and sensitivity to pain; and many other internal senses.

Information is better remembered when accompanied by a visual image. Many would agree that visual imagination makes one's mind more creative and resourceful. Much attention goes to intuitive analysis in medicine, which may go beyond the analysis of numerical data coming from instruments. Intuition combined with good logistics enables us to envision the opponent's future moves, which is a valuable ability in chess and other strategic situations. We do not need the physical presence or images of physical objects to imagine them vividly. Visual imagination means we can envisage real or imaginary content and present it using signs, icons, symbols, metaphors, and analogies, sometimes using mental shortcuts to understand various concepts.

André Malraux (1958/1974, 1996) introduced the concept of the museum of imagination, the imaginary museum without walls – a collection of artworks held in a person's memory as essential and favorite. It is a collection of previous experiences, visual references, and notions developed in a viewer's mind, which may enrich the viewer's reception of artworks and the art–viewer communication. We may access any exhibition online and advance our visual memory and imagination.

Activity: Sketching feelings and emotions

Find a hole among the clouds. Imagine yourself hearing, feeling, and sensing a heavy rain after a hot noontime. See in the mind's eye the physicality of the raindrops, their contact relating to your body rather than your mind. Make a very short sketch showing you in such a summer shower and the excitement it caused.

Problem-solving abilities and intuitive decision-making abilities hinge, in some measure, on imaginative thinking. More than teaching the game rules is needed when teaching to play chess. An imaginative approach to a task of this type involves developing intuition that allows finding effortlessly optimal thoughts and solutions (even if they are not justified). Intuitive processes support our aesthetic preferences and may lead to scientific discoveries.

Imagination is instrumental in digitally created stories, animations, and filming techniques, making Superman or Batman's impossible actions look real. In turn, they may help to develop the imagination of viewers. However, accomplishments of precomputer works often surpass these productions, for example, *Gulliver's Travels*, 1726, by Jonathan Swift, *Frankenstein*, 1818, by Mary Shelley, *The Lord of the Rings* series, 1954–55, by John Ronald Reuel Tolkien, and fictional characters such as Count Dracula, Nosferatu, Dr Jekyll and Mr Hyde (eponymous novel published in 1886 by Robert Louis Stevenson), or Dorian Gray, (*The Picture of Dorian Gray*, 1891, by Oscar Wilde). Unsurprisingly, authors of comic books, games, animations, and movie scenarios remake these works in many ways, with results that may overshadow many other attempts.

Activity: Imaginative watching

Look out the window. Count how many people you can see. Find names that fit and assign a name for each one of them. If no people are present, imagine two people talking about what is in front of their window and then write a few sentences they speak.

Project 7. Impossible creatures

This project invites learners to picture and describe imaginative characters as nonexistent creatures. These characters may be used later in other projects, for example, as characters in animations.

Getting acquainted with impossible creatures

Our imagination may be rooted in nature. For example, we can see pictures of a minotaur – half man and half bull – or a unicorn – a horse with a single horn projecting from its forehead. Images of impossible beings may present unusual types of motion when applied to imaginative creations, a type of movement typical of whales, butterflies, birds, quadrupeds (having four legs), or humans (Figure 4.2). Also, differences among organs in various species may inspire you, such as bronchia for some fish, bronchi in humans, or gills. Bio-inspired inhabitants of this planet often become presented as aliens.

We can find imaginary examples in ancient Greek religions, which included gods and half-gods. Each one had powers and weaknesses mirroring the earthly

Figure 4.2 Impossible characters

human and unnatural forces. We can also find imaginary bio-inspired creatures in literary works, sometimes coming from translations from old languages. For example, *Beowulf* is an epic heroic poem written c. 700–1000 AD in a West Saxon dialect of Old English. In 1967, the Argentine writer Jorge Luis Borges (2006) assembled an imaginative bestiary of various actual, fantastic, or mythical animals in *The Book of Imaginary Beings*. In his book *Bestiary*, Nicholas Christopher (2008) discussed the fate of animals that missed Noah's Ark. He described and sketched cross-breeds such as a werewolf and other harmful and quite unfriendly creatures. Caspar Henderson (2012) took a different approach: in his *The Book of Barely Imagined Beings*, he describes the real animals that are stranger than the imaginary ones, adding his philosophical ruminations along with evolutionary and social comments.

Impossible characters enhance connections between biology, engineering, and material sciences, so impossible creatures are becoming possible and quite realistic. There is a growing partnership among academia, laboratories, and industry aimed at creating new materials and structures. Production of new materials is often inspired by fantastical beings and then made by the translation of biological data to materials science. To apply bio-mimicry, people are watching living organisms and designing new structures and materials that function and respond to external stimuli in a way similar to living beings. With bio-inspiration, systems inspired by structures in organisms that exist in nature include optical fibers, liquid crystals, or structures that scatter light. Bio-derivation means the application of existing biomaterials to create a hybrid with artificial material.

While looking at impossible beings and designing your creations, place your genie among the impossibilities, a spirit who would grant your wishes when you call for him. What would be the two wishes for your genie? These requests should go beyond what you can find yourself (love, friendship, or better habits). Also, do not ask for info you may find on the internet. For example, you may ask, "show me the sixth dimension" or "explain my protective powers."

Activity: Sketching impossible creatures

Imagine anything that flies: a bird, a bat, a butterfly, or anything else. Assume that some impossible creatures like to travel on these animals as passengers. What would these impossible flyers look like? What type of symbiosis would you envision being present among them?

Creative thinking

According to Reche and Perfectti (2020) and Scheffer (2014), creativity is an evolved cognitive mechanism for abstracting, synthesizing, and solving nonrecurrent problems and is crucial for performing cutting-edge science. Developing creative thinking aims to stimulate curiosity, make novel creative tasks, and promote divergence – differences in opinions, interests, or attitudes. Creativeness may involve imaginative, associative, metaphorical thinking and originality, fluency, elaboration, brainstorming, and collaboration when working in a group. Creativity means that one is open to changes and novelty, willing to accept new possibilities or others' points of view, is flexible in attitudes to life, and appreciates good work. Creative thinking uses imagination to "think out of the box" and beyond the routine. In architecture, we may see many unique, original solutions in designing new buildings, using new materials and technologies and planning landscape architecture of outdoor public areas. For example, Maya Lin, the author of the Vietnam Veterans Memorial in Washington, DC, traces the lines of the natural landscape and, with the minimalist approach to public art, creates architectural structures as swells of earth that belong to their surroundings. In her indoor sculptures, she focuses on environmental issues.

Creative thinking skills can be seen in many children and adults, not exclusively in inventors, artists, musicians, and writers. Creativity can be enhanced at any age and knowledge or skill level. Innovative problem-solving allows analyzing the facts, applying analytical skills, and making connections between seemingly unrelated concepts. Creative thinking processes can be convergent when one uses deductive reasoning to analyze scientific evidence and find the best solution to a problem. Divergent thinking prefers inductive reasoning to explore many possible ideas and find solutions spontaneously in a nonlinear way. Divergent thinking on idea generation and convergent thinking on idea analysis support the creative thinking process.

Former President Obama announced that creativity is a currency of the 21st century. Creativity is essential for doing scientific research and making innovative inventions and unusual solutions. However, students are trained primarily on critical, logical, analytical, and evidence-based convergent thinking, and rote memorization of the accumulated data may block their creative thinking abilities. There is a shift in the arts to serve social, moral, or political issues that replace beauty and poetics. Students often draw inspiration from existing works and AI images. Creative thinking may help preserve beauty in original creations.

Mihaly Csikszentmihalyi (1996), who studied how creative people live and work, stated that creativity is a central source of meaning in our lives, and most things that are interesting, important, and human are the result of creativity. He introduced the concept of flow, a creative, focused mental state (Csikszentmihalyi, 1997, 1998). Learners' creative development makes it possible to enhance the capability of problem-solving and problem-finding, the ability to focus on a creative task and perform in full flow (Boden, 2009, 2012).

Thinking creatively allows you to create, invent, problem-solve, and communicate successfully. By developing creativity, you may become an expert in your area, understand complicated problems, and invent novel or innovative solutions to these problems. Curiosity – eagerness to know and learn about novel things, the inquisitive interest in new ideas, creations, and discoveries – strongly supports creative thinking skills. NASA named its Mars science laboratory rover *Curiosity*. Curiosity and taking risks driven by curiosity may fight tiredness and boredom.

Activity: Practicality and appearance

Think about the relations in architectural design between functions required to serve a purpose well and the everyday elegance of the building – being graceful and stylish in appearance. Find a noticeable building in your neighborhood. Sketch a quick draft of the building's features and then exaggerate its importance and prominence by adding some fine details to its structure.

Digital creativity

Digital creativity involves cross-disciplinary collaborative efforts. Even if someone creates alone, the technology involved in the process has already been developed by somebody else. This trend is intensified by using artificial reality as a part of creative works, as it collects data by copying, rephrasing, or mimicking other people's works. Collaboration became possible online with participating individuals or groups working on different continents since the web inventor and director Tim Berners-Lee initiated the World Wide Web Consortium (W3C), which developed international standards for the web: HTML, CSS, and many more. You can use the W3C WAI resources to make websites, applications, and other digital creations accessible and usable to everyone (W3C Web Accessibility Initiative (WAI/) (2023).

Art creators are working on their projects beyond the left-brain, right-brain approach to thinking and making art products a part of science- and technology-related solutions, presented as pictures and interactive and interdisciplinary visual forms of storytelling. The film director David Lynch supposedly said that negativity is the main enemy of creativity (Reche & Perfectti, 2020). It is vital to reduce students' anxiety, support them in developing their creative potential, and provide them with tasks and projects they can enjoy.

Collective creativity

Collective creativity may grow from exchanging ideas with people who create a network of minds. A diverse, nonhierarchical team can best develop collective creativity. Brainstorming sessions, communication between coworkers, and good cooperation among colleagues are most important for increasing collective creativity (Reche & Perfectti, 2020).

This book encourages readers to create new action-based stories designed for specific audiences. Learners are invited to seek their own solutions, create their own messages to the world, and avoid copying anyone and anything. Be you. Do not compare yourself to anybody. Develop your own interpretations, outlook, views, and creations. When someone becomes competitive because of comparing, such a person begins to dislike and stop all that is above and beyond their possibilities. This attitude cuts this envious person from the best people.

Metaphorical thinking

A metaphor reflects cognitive operations; it makes us see one thing in terms of another and transfers meaning from one thing to another. Conceptual metaphors allow for understanding an abstract or unfamiliar domain in terms of another, more familiar concrete domain (Lakoff, 1990; Lakoff & Núñez, 2001). With a metaphor, we relate complex or unknown concepts that are not easy to understand to things we already know.

Communication goes on with the use of language. Our thoughts and language are highly metaphorical. Metaphors are not true or false. They are present in our thoughts; thus, we can perceive invisible concepts. Metaphors, especially biologically inspired metaphors, are widely used in computing, data and information visualization, data mining, and the semantic web.

Visual metaphors

In large part, people think in pictures. Metaphors are often visual. Images serve as metaphors, icons, and archetypes. The symbol of 'heart' is a metaphor. Visual metaphors help convey abstract concepts and reinforce learning. They can be found everywhere, with images of children as a metaphor for a school crossing, a person in a wheelchair for parking for disabled people, and a letter P for a car parking place. However, these symbols are not always used consistently; symbols used on maps include a letter H for a hotel and a red cross for a hospital, while the road symbols are a letter H for a hospital, a house or a bed for a hotel, and a red cross for an emergency. Typical clipart and clipart included in writing software are metaphors. Terms such as 'swarm computing' and 'cloud computing' and the design of links on the web all originate from biology-inspired metaphorical thinking.

In many instances, art is metaphorical. While photographs show the real appearance of the object under inspection, metaphorical visualization helps us to understand the invisible factors in the conceptual, geometrical, or literary arrangements. A double helix metaphor helps us grasp the essence of the protein structure.

Metaphors in visualization

We may use the same metaphor to visualize many concepts. In visualizations, ideas that are pictorial and linguistic at the same time are presented as metaphors. Somebody studying more than one field of knowledge may present a complex concept using a word or an image belonging to the other field. Also, natural forms serve to design patterns in ornaments, architectural details, and interior design projects. In many cases, scientists create biologically inspired metaphors. Metaphors derived from nature often serve for developing computing methods, such as artificial neuronal networks, evolutionary algorithms, swarm intelligence, genetic engineering techniques, and bio-inspired hardware systems.

The sensation experienced with one sense, for example, sight, may serve as a metaphor for a process involving another one. The poem below presents how it can happen.

Shifted Senses

we look at music when attending a concert,
we read artist's statements when visiting a gallery,
we talk about food at the dinner table,
we watch literature in a movie theater,
we taste drinks to appreciate their smells,
we touch the screen to Like on social media,
instead of patting you on your shoulder.
we use senses we do not know we possess,
to talk about ourselves,
not sensing our surroundings or growing our curiosity.
we see art with no elements and principles of design in them,
yet we create new venues and outlets for our new experiences.

Bio-inspired metaphors for technical or scientific notions are often mimicking living systems and just evoke responses typical of our reactions to natural objects or events. Many times, a metaphor originates from an inspiring biological structure. Also, many technical terms resemble biological ones. For example, we use a mouse to navigate or draw on a computer screen.

A tree metaphor is most frequently applied in computing, business, and other domains. For example, a tree serves as a metaphor to analyze hierarchical structures in which each segment of a structure (except the highest-ranking one) has an inferior position concerning a higher segment. T. Manuel Lima called the tree figure the most ubiquitous and long-lasting visual metaphor "through which we can observe the evolution of human consciousness, ideology, culture, and society "(Lima, 2014, p. 42). Around the world and over the ages, information design presents data placed in circles and uses forms present in nature or human environs (Lima, 2017).

Semiotics

The main object of semiotics is the meaning of our thoughts and actions. Semiotics examines the meaning of signs, sign systems, symbols, and the relations between signs and the things to which they refer. The sign systems may include natural languages, writing systems, musical notations, garment codes, culinary codes, traffic signs, and even film as a system of signs. Animals give a symbolic meaning to things they see or hear. For example, the rustle of leaves may mean danger for small animals.

Visual semiotics

Visual semiotics studies communication through images using signs, visual metaphors, visualization techniques, symbols, analogies, icons, allegories, and time-based images. The same signs may mean something different in another country; we may not even know we convey strong messages. Signage determines a design by using signs and symbols, for example, on billboards, posters, or street signs. Also, dance, mime (performance art involving body motions and facial expressions without speech), and pantomime (a musical theatrical production using gestures and movements but not words) convey meaning, often linking haptic (related to the sense of touch) and visual modes.

Intelligence

It is still unclear why people differ in intelligence (Deary et al., 2022) and which factors decide individual intellectual abilities. People hone their intelligence when they go through the cognitive, sociocultural, psychological, and other developmental stages. Intelligence is usually ascribed to one's ability to learn, understand, adapt, and manipulate new information. Intelligence is needed to understand abstractions with no physical features, for example, in art, computing, mathematics, physics, logic, or information processing. Abstract reasoning about complex, symbolic, or hypothetical concepts, ideas, principles, and experiences involves intelligence. Without an adequate level of intelligence, abstract concepts such as wisdom, humor and jokes, freedom, jealousy, and imagination are interpreted only on the concrete level. The same can be said about acquiring and applying knowledge and skills. We use intelligence when we explain something figuratively or through a metaphor, explore patterns and relationships in objects and events, or build a theory that explains something. Quite often, a student who was labeled at school as less intelligent is the one coming up with the best inventions.

Collective intelligence

An emerging increase in the interactions among cooperating teams of colleagues from different locations or institutions resulted in a growth of interest in collective intelligence as a factor of progress. Collective intelligence does not depend on the presence of brilliant participants or the average intelligence of the team members. It seems that the growth of collective intelligence is supported by the social skills of the team members and their habit of speaking in turns within a group. Communication networks, not exclusively face-to-face conversations, support the growth of collective intelligence. Hierarchical groups, in which one or two people monopolize communication, and it goes in a top-down direction, will have low collective intelligence and a low number of interactions. Generating productive networks nationally and internationally supports the development of collective intelligence (Reche & Perfectti, 2020). Diverse teams recruiting from different groups of age, race, gender, and affiliation can grow higher collective intelligence than homologous groups that do not show various perspectives and skills. However, the growth of higher collective intelligence may happen when the group diversity does not provide language or cultural difficulties.

Space and spatial abilities

Space

Space is the void between solid objects and shapes, which is present everywhere around us. The visual arts are considered the space arts or spatial arts because they take up space, two dimensions in the case of drawings, paintings, and prints, or three-dimensional space taken up by a sculpture and architecture. Music and literature take up time. Some arts, such as film, opera, dance, and theater, take both space and time. Space taken up by solid shapes and forms is called positive space. The negative space describes the void between shapes and forms in a design, picture, sculpture, or building. A doughnut has a positive shape or form; the doughnut hole is a negative shape or space but becomes a positive one when served baked.

Sensory stimuli evoke specific reactions in our cells or whole organs. Our brain transforms sensory stimuli into perceptions. In response to spatial perception, our brain performs spatial cognition, which compares patterns gained through sensory stimuli with our memory representation.

Spatial abilities

Our spatial abilities may be measured by our competence in spatial visualization (an ability to mentally rotate, twist, or invert pictorially presented objects) and spatial orientation (understanding the array of objects, utilizing our body orientation when we perceive objects). Spatial cognition allows us to reflect and reconstruct the space in our thought based on our perception (what is seen when we process sensory information).

Spatial skills are necessary for many activities and occupations because they allow the use of visualization – explaining something visually to oneself from the viewpoint of one's mind's eye. Spatial visualization and orientation skills are measured through reliable and valid tests. There is a high correlation between spatial skills and creativity. The reliable spatial tests measure spatial skills and spatial visualization abilities.

When we apply visual and spatial thinking to our learning, we may produce meanings and connotations that cannot be achieved using language alone. Verbal, mathematical, and spatial reasoning are the specific cognitive aptitudes considered predictors of future achievement. Also, thinking in 3D can improve math and science skills (Wai et al., 2022). For all these reasons, teaching students how to create graphics using 3D software would be recommended.

Conclusion

Cognitive, creative, metaphorical, and psychological processes play an essential role in visual communication. They are both numerous and diverse. This chapter describes mental capacities that enhance visual communication and support the creative collaboration of professionals with learners and their ability to produce successful visual projects. Actions toward improving cognitive capabilities depend on one's literacy level – knowledge or skills in a specified area. Thus, literacy may be digital, visual, cultural, scientific, technological, computing, media, or competence in other areas. Projects and activities that can arise attentiveness, curiosity, and motivation support the growth of corresponding literacy.

5 Techniques and strategies for including novelty in instruction

Introduction

This chapter discusses components that are important in visual instruction and learning. The first part describes concepts and techniques supporting the visual transfer of knowledge that shapes our living realm. Description follows of the data-, information-, knowledge visualization, and serious games. The second part is about strategies supporting introducing novelty to instruction, including the STEAM, STREAM, and STEM programs and copyright-related issues. The chapter comprises eight learning projects related to the themes under discussion.

Concepts and technologies supporting the visual transfer of knowledge

Artificial intelligence and the novelty brought about by AI techniques

Computers can, with the use of AI-based programs, perform many tasks that till now were done by humans: they conduct research much better than previous search engines and produce texts, summaries, notes, poetry, and much more, evolving and improving every day unpredictably.

NATURAL LANGUAGE GENERATION AND NATURAL LANGUAGE PROCESSING

AI tools cannot create texts in an organic way, like humans using natural languages. Large language models (LLMs) are computer programs for natural language processing. They use artificial neural networks to generate text. Examples of the large language models are GPT-3, GPT-4, or a conversational generative AI chatbot Google Bard. Large language models enable making AI chatbots, AI search engines, and other AI applications.

DOI: 10.4324/9781032705347-5

To tell a story in human language, one has to use two kinds of AI software: natural language generation and natural language processing. They generate and process natural language according to the user's prompts and description. First, the natural language generator (NLG) gathers a large amount of data to compose understandable texts in the human language, using machine learning to analyze the text or speech data (IBM, 2023). For example, it can extract information from an Excel document and produce the content in a few seconds. Then, the following AI tool, natural language processing (NLP), interacts between the computer and human languages so that the computer can process and analyze natural language data.

The AI-based technology can provide a content summary of each book, performance, or idea. However, we risk losing some level of beauty offered by talented authors. Students are sometimes asked to design a book cover using a brief plot summary as a base for this task. The whole area of imaginative narration, poetic descriptions of people, places, events, distinctive tone, suspense, and expectations are lost for these students. Similarly, the same values will never reach those designers who limit themselves to using only the facts described by a bot. Often, one can tell that a text is generated by AI when there are repeated words and phrases and unverified data and the content is unrefined below academic standards.

Examples of AI-based accomplishments   AI-based software and applications such as OpenAI, GPT, AI ChatGPT-2, AI ChatGPT-3, AI ChatGPT-4, and more can read people's minds (Dazed, 2023), recreate with AI images a person has seen (Nahas, 2023), and offer embodied artificial intelligence (Holoworld, Artificial Intelligence, 2023). The ChatGPT-4 machine can, equally well as humans or even better, perform coding and programming, software and web development, graphic design, content and technical writing, customer support, translations, business and market analysis, and many more tasks (Mok & Zinkula, 2023).

Batch processing enables carrying out jobs submitted by users as software programs without interactive or manual intervention. Batch programs can run automatically. Automatic equipment enables conducting a fast-increasing number of processes.

The brain scan to text translates the brain responses to a speech into understandable language. As a result, it is possible to read another human's thoughts for the first time. The researchers examined the fMRI data to evaluate the encoding model that predicted the brain responses to a speech test story in most cortical regions outside of primary sensory and motor areas (Tang et al., 2023).

The text-to-image apps make this procedure. For example, Midjourney's Photoleap AI Image Generator, which is also a photo editing app, applies a text-to-image technique to generate AI images based on clients' descriptions of imagined pictures and makes graphic editing of photos.

Image to image. In 2022, scientists trained an AI machine to transfer brain scans to images: to recreate images previously seen by people. Scientists from Osaka University in Japan developed a Stable Diffusion model, which made it possible to reconstruct images from human brain activity obtained via functional magnetic resonance imaging (fMRI). The OpenAI DALL-E 2 created imagery based on the text input. With latent diffusion models, researchers reconstructed images from the human brain activity of image perception areas. Humans saw thousands of images, their brains decoded images, and the fMRI picked up blood flow changes and detected oxygen molecules close to neurons. With AI, the authors reconstructed images from the fMRI recordings (Takagi & Nishimoto, 2023). This technique will possibly serve to communicate with paralyzed patients who cannot speak or write. Kneeland et al. (2023) decoded human brain activity of the visual cortex to examine images conditioned on this descriptor. They selected the images that best predicted brain activity and obtained high-quality reconstructions of images.

The brain images scans to art. In experiments called neuro-art, AI uses fMRI scan data to make artistic pictures. Neurographic art examines how the arts and aesthetic experiences change the brain, body, and behavior (Drevitch, 2022).

The brain activity into language. People listened to recorded speech read aloud, and their brain waves were recorded using magnetoencephalography (MEG) or electroencephalography (EEG). The AI studied these 169 people's brain waves, looking for patterns between the listening to audio recordings and the brain activity of people listening to the recordings. With the MEG data, the sound recordings matched the brainwaves with 73% accuracy (Sullivan, 2022).

More new solutions are possible such as text-to-video, text-to-PPT, text-to-voice or speech, or text-to-music. They are published almost daily, and it is a challenge to tell how many types of AI there are. People compare AI and AI-enabled machines to human brains and then classify machines according to their likeness to the human mind. The AI or AI-based systems types include reactive machines and limited memory machines. AI specialists study the theory of mind and the self-awareness of AI. However, these groups are tentative because, as yet, AI-based machines cannot think and feel like humans.

The AI GPT-4 can read the sensory and motor areas in cortical regions (Tang et al., 2023).

However, deep learning and AI GPT-4 do not yet have consciousness, intuition, or self-identity. Intuition is considered the highest level of consciousness. Maybe intuition and creative thinking are not verbal, so deep learning may not read our deep thoughts but only our verbalized perceptions and images.

Machine learning, a subset of artificial intelligence, is the capability of a machine to imitate intelligent human behavior using computer systems to learn without instructions but using algorithms and statistical models. Machine learning imitates and makes more precise the way that humans learn. Artificial intelligence systems work similarly to human problem-solving capabilities (*https://mitsloan.mit.edu/ideas-made-to-matter/machine-learning-explained*). Machine learning systems learn from the data and adapt themselves according to them. They enable the creation of intelligent environments. For example, portable 3D visualizations published on the web or offered wirelessly are used for marketing events, sales presentations, and product animations.

Project 1. Visualize cooperation on novelty

In this project, visually present a cooperative work on some novel technique. Choose a new technology that is somehow related to your current or future career or aroused your attention and interest. For example, find a biology-inspired technology based on social animals living within their shared habitat. Examine swarming of insects, flocking of birds, herding of the four-legged animals, or schooling of fish. Some animals live in colonies and strongly interact among themselves and with their surroundings, for example, penguins, bees, termites, and ants. Colonies of insects live in social structures and have castes with different responsibilities. Individual actions give them substantial collective power, so maybe, in your visual story, they could serve as a metaphor for cooperative striving toward a novel invention.

Preparation

When you have already decided what technology you will choose to learn about, gather information about its practical applications. Beyond doubt, developing this technology involved the cooperation of people with different areas of specialization. In the next chapter, you will learn about people from various science areas involved in cooperation, their equipment, and novel technology applications in everyday life. You may discover that people who developed new methods or materials have often been bio-inspired by animals coping successfully with challenges these researchers had to conquer. For example, biologically inspired computer techniques include artificial life, artificial neural networks, swarm intelligence, and other artificial intelligence techniques.

Researchers and engineers use biological materials or functions as bioinspiration for creating similar ones but from different materials. For example, they look at the gecko's adhesive foot: it can work in a vacuum and underwater, is self-cleaning without any residue, and adhesion is reversible, so when it runs up walls, a gecko sticks and unsticks itself 15 times per second (National Research Council of the National Academies, 2008). Thus, scientists work on creating self-evolving, self-healing, self-cleaning, and self-replicating supermaterials. They also strive to mimic naturally existing structures that can evolve and adapt. Think in these terms about machine learning and the domain of artificial intelligence. Currently, researchers develop advanced materials with integrated electronics. They work on applying the direct laser writing in the living neurons to 3D print micrometer-size flexible conducting polymers and then stimulate the brain tissue (Baldock et al., 2023). It will allow monitoring in real-time medical procedures, e.g., against epilepsy or pain.

Make your plan of action and the story you will tell, and think about a container (see p. 16 of this chapter) you will use for this story. You may make your work as an animation, video, comics, presentation of sketches with captions, or in any medium you choose.

Working on a project

Once you decide on the project's theme – the technology you will describe and the story you will tell – create a metaphor for the cooperation of people working toward a novel invention. Pick an animal from those that live in colonies, communities, or groups and have complex communication systems. Start sketching such group members while running, carrying out various actions, or resting. Make many quick sketches so you can choose from varied, diverse material. Figure 5.1 shows sketches of black-hooded rats nesting at home as a social group. Rats have a social hierarchy. They are known as helpful to each other just as humans are, especially those who have helped them; this is unusual among nonhuman beings. They cooperate by giving each other food, grooming other rats they like, and can catch a scent of cooperative actions, which triggers their altruism.

While choosing a species for your story, try to match the animals' traits and way of living with the technology and people working on it. For example, as swift and active animals, rats can serve as a metaphor for a group developing dynamic, fast-changing techniques such as fiber lasers (see what follows) or laser-assisted welding (see Chapter 6).

When sketching, experiment with different tools (Figure 5.2). Name individual characters you selected from your sketches and assign different functions to creatures as they contribute to the common task. If needed, your animal

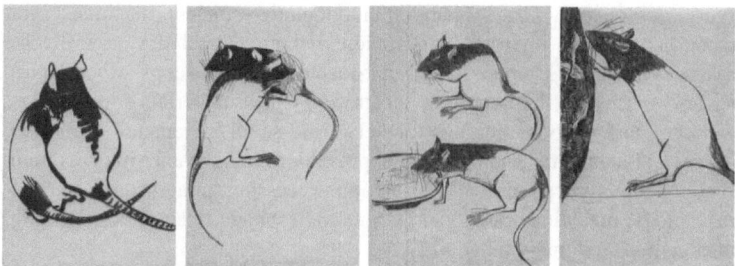

Figure 5.1 A company of black-hooded rats

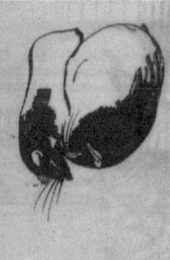

Figure 5.2 A company of rats; sketches in pencil, pastel, and ink

characters may get some help from other parts of the animal community so that you will demonstrate their cooperative work on a common goal.

Connect your sketches with informative captions. Tell about developing the technology you choose to demonstrate through the animal metaphor; all presented as the story you will tell and show. Design your whole work as a comic, animation, or video with the characters sketched by you. You may want to convey information about technical details through fictional dialogues among your characters. Depending on the theme of your work, add some schematic sketches of the background supporting your project.

More technologies used in transferring knowledge visually

Photonics is the study of generating and harnessing light and other forms of radiant energy. Beginning with the ancient cultures worshiping light, people study the properties and transmission of photons. Photon is a quantum unit, a particle being a unit quantity of light or other electromagnetic radiation. Photonics comprises technologies that are becoming vital in our current circumstances. Leading technologies include imaging techniques (e.g., for microsurgery), lasers (e.g., lidar systems in autonomous vehicles), machine vision (humans and robots interpret information without optical devices), precise measurement devices, optics and super-resolution microscopy, fiber optics, fiber lasers, sensors, light sources, spectroscopy, new materials and coatings, positioning and nanopositioning (for delivering small increments in motion, up to 1/10 of a nanometer), and more. Applications include energy generation, detection, communication, and information processing, in aerospace, agriculture, biology, medicine, clean energy and environment, electronics, lighting, transportation, and many other areas.

Biophotonics brings light to the life sciences. It means photonics products and techniques help to solve problems for researchers, product developers, clinical users, physicians, and others in medicine, biology, and biotechnology (BioPhotonics, 2022).

Optical science is related to light and vision, primarily to visible, infrared light and ultraviolet, but also other types of electromagnetic waves such as electromagnetic radiation, microwaves, X-rays, and radio waves. Applications include generating data in the IoT (Internet of Things) network, transporting and routing data, information technology, telecommunications, health care and life sciences, remote sensing, lighting and energy, national defense, industrial manufacturing, manufacturing optical systems, optics research, education, and optical sensing and imaging (Kyriacou, 2019).

Lasers emit photons from excited atoms or molecules to generate mono-chromatic light or other electromagnetic radiation. Lasers range from a size of a quantum dot (a nanoscale particle of semiconducting material) to football-field size. Laser technologies involve CO_2 lasers, infrared lasers, and fiber lasers that use fiber optic cables (see below) and are faster, stronger, and cheaper than conventional ones. They can cut stones, metals, glass, and most other materials with a precision of about 1/4 of human hair. Applications include cutting, drilling, and marking objects and medical uses of lasers and microlasers in surgery and clinical practice. In optical science, semiconductor lasers are used in single photon laser scanning microscopy and optical telecommunication using infrared light; lasers are also used in holography, reading bar codes, recording and playing compact discs, and other applications.

Fiber optic cables deliver data, voice, video, and images when the light goes through hair-thin, mostly glass fibers. Fiber internet delivers broadcast fast through fiber-optic cables, which do not use electric current like older internet but light signals going through the fiberglass cables. Fiber optics link computers in networks, examine human internal organs, and control manufactured products in industrial, avionic, military, and communications systems. Optical fibers allow sensing physical signals (temperature, pressure, displacement, vibration, acceleration, electrical field, chemical compounds, and more). They serve to deliver power and illumination.

Quantum physics and quantum computing, which are now in a phase of fast development, will outperform the capabilities of traditional computers in many areas, primarily regarding their speed to efficiency. Quantum technologies, including silicon nanophotonics, mimic chemical and biological systems and strengthen the existing science and technology productions. Optogenetics controls living cells, mostly neurons. Plasmonics studies quantum units of plasma using photonics methods.

The economic impact of quantum computing will be most substantial in the automotive, chemical, financial services, and life sciences areas. The Quantum Computing Network promotes collaboration with academic institutions, national governments, and big companies such as Amazon, IBM, Microsoft, and more. The quantum computing industry, which has become prominent, needs workers with multidisciplinary skills. The same demand comes from the fields of photonics, biophotonics, and optical sciences and technologies.

Project 2. A timeline of collaborative inventions

This project aims to show how inventions in various areas of knowledge have been developed in different times and places, some occurring at the same time or in the same region. To get such a general perspective, building a timeline seems useful. The further text provides the learners with detailed guidelines for building this project.

The first step is data collection. Gather information provided in this chapter and support information with facts you can find online. Add descriptions of inventions included in other chapters.

The next step will involve making quick sketches denoting inventions and those who made them. Make these sketches simple, showing only the essence of the theme, just as described in Chapters 1 and 4. Also, design icons for different sciences and place them around the sketches to emphasize the collaborative nature of the work ending with a successful invention.

After that, visualize the times of these events. For this purpose, build a timeline showing the centuries, starting from prehistoric times, then the ancient period, the Middle Ages, Renaissance, and up to the present. Write a specific date or the approximate period when a particular invention was made. Typically, the x-axis denotes time.

Develop a color-coding arrangement to assign distinct cooperative inventions to specific places on the timeline. Allocate the inventions to countries where they were developed, and assign colors to those countries. Many times, cooperative work was done in many places at the same time. For example, laboratories in many countries were involved in developing a CRISPR technique (including crystallography specialists). That means several colors will adorn the items, denoting inventions that contribute to developing one novel technique.

Write a caption explaining your sketches, icons, and color coding for each item on your timeline.

Data, information, and knowledge visualization

We can collect data as numerical – presented as numbers or categorical data that belong to particular categories. Categorical data include names, locations, web addresses, emails, file paths, shapes, colors, or emotions. We can compare numerical data with their categorical matches. Numerical and many kinds of categorical data can be presented as visual objects: flat, tridimensional, time-based, or virtual. They take the form of pictures, points, lines, graphics, or animations. Many data visualization techniques offer, for example, bar charts, pie charts, tables, line graphs, diagrams, animated graphics, and many more. Different kinds of software can be downloaded, often for free, to visualize data. Infographics (graphical presentations of data, information, or knowledge) visualize ideas and concepts online, in newspapers, and in other publications. Social media (e.g., LinkedIn – a business and employment-focused social media platform – Facebook, or X, formerly known as Twitter) became popular as fast tools that are convenient for people with a short attention span.

Visualizations often offer many kinds of virtual reality technologies that use computer modeling and simulation to allow users to interact with an artificial environment. **Virtual reality (VR)** provides complete immersion in the virtual world with a 360° video of real surroundings and/or computer-generated synthetic images. **Augmented reality (AR)** spreads, without occlusion, computer-generated imagery over the real content so it looks like interacting with the real environment. For example, when you examine furniture in an IKEA store with the mobile app IKEA Place, you can use augmented reality: with the camera on your mobile device, place digital furniture anywhere at home and move or rotate it within view. You can virtually see on your phone whether it has the right size for your room. **Mixed reality (MR)** overlays synthetic images on real ones so they occlude and interact in real time with real objects. For example, instruction can flow around parts of a machine to be fixed. **Extended reality (XR)** refers to all environments generated by computers that mix reality and virtual worlds, including wearables (Irvine, 2023).

Specialists are working on integrating visualization and AI techniques to apply them to complex data analysis. Processing big data requires complementing the strengths of visualization and AI. Scientists work on the levels of integration, first in one direction: VIS > AI or AI > VIS, and then bilateral: VIS + AI, which will provide barrier-free communication between human and artificial intelligence (Higher Education Press, 2023).

Data visualization means creating graphical representations of information (Miller, 2019), thus translating a concept or idea into visuals. People create visualizations to understand abstract concepts better. When we use an image or a chart to convey a message or explain something, we are making a visualization of our message.

Prehistoric cave paintings, Egyptian hieroglyphs, or geometric figures drawn by ancient Greeks are early forms of visualization. Currently, visualizations support our work in various areas of activity. For example, Harvard Business School Online presents illustrated descriptions of data visualization techniques that help make data-driven decisions (Miller, 2019).

To see examples of visualizations, visit Data Is Beautiful: 10 Of The Best Data Visualization Examples From History & Today (*www.tableau.com/ learn/articles/best-beautiful-data-visualization-examples*). Go to The 30 Best Data Visualizations of 2023 (*https://visme.co/blog/best-data-visualizations/*). Visit Visualizing the History of Pandemic (LePan, 2020, *www.weforum.org/ agenda/2020/03/a-visual-history-of-pandemics*).

Information Is Beautiful by David Mc Candless (*www.amazon.com/ Information-Beautiful-New-David-McCandless/dp/0007492898/ref=sr_1_1?c rid=1TIC41QBOJS96&keywords=Information+is+beautiful&qid=16878897 00&sprefix=information+is+beautiful%2Caps%2C143&sr=8–1)*

Knowledge Is Beautiful by David Mc Candless (*www.amazon.com/ Knowledge-Beautiful-Impossible-Invisible-Connections-Visualized/ dp/0062188224/ref=sr_1_2?crid=1TIC41QBOJS96&keywords=Information +is+beautiful&qid=1687889732&sprefix=information+is+beautiful%2Caps %2C143&sr=8–2)*

Information visualization presents data in a meaningful, visual way so viewers, often nonspecialists, can easily understand them. When presented without context, raw data alone does not provide meaningful input. With information visualization, pictures of objects are connected through their relations; data are defined, selected, transformed, and represented. Visual presentation of abstract data supports human cognition and allows users to draw insightful conclusions.

Interactive information visualizations allow users to manipulate and apply their content according to their particular conditions. Interactive visualizations give users control over what they want to see. They can also narrow the data down to their personal interest, such as age, place, or another factor. We can find examples of interactive information visualization when we visit a weather channel presenting graphs and charts or use services such as Google Maps or media services like Netflix (Gloat, 2023). Internet users can see 3D objects on the web. An interactive Periodic Table of the Elements can be found at *http://chemicalelements.com/elements/rf.html*

Visualizations may be presented in many file formats depending on the visualization type. For example, illustrated presentations or infographics are delivered as pdf (portable document format), ppt (PowerPoint presentation), jpg or tiff image, png (portable network graphic), dashboard (graphical summary), or gif image (supporting animated or statistical images). Interactive data visualizations are presented in JavaScript language, CSS, HTML (Hypertext Markup Language), and more. AI delivers alternative ways to visualize your ideas. Animated data videos may take a format of mov (Apple quick time movie file), mp4 (digital multimedia container format), aep file, a video project created by Adobe After Effects, a special-effects video-editing application with the filename extension. aep or. aepx), and many more. Perhaps visualizations offered by NASA (2023) belong to most captivating images and videos, for example, NASA Image Galleries (2023).

Programming languages are used for working on complicated data, such as analyses of statistical data or big data (which are large, hard-to-manage data). For example, Python is a general-purpose programming language. Python Bokeh is a library for creating interactive visualizations for web browsers. It provides interactive charts and plots. Python Bokeh uses HTML and JavaScript to present the construction of novel graphics with high-level interactivity

(GeeksforGeeks, 2022). The R programming language serves for statistical computing and graphics. It is used for storing, manipulating, and retrieving data in databases.

[The app *Spacecraft AR* uses augmented reality technology to put virtual 3D models of, for example, COBE's (2023) robotic space explorers into any environment with a flat surface (*www.nasa.gov/sites/default/files/thumbnails/image/ar20180320-16.jpg*). Earth overview is at *https://solarsystem.nasa.gov/planets/earth/overview/*.

Knowledge visualization enables visual communication between individuals through images, graphics, animations, and moving images, using them to create, integrate, and apply knowledge rather than data. Knowledge visualization presents the images to support learning in an engaging way, to understand better and improve recall. Specialists convey information in education, journalism, business management, technical production, and numerous other professions. Explaining various concepts and events across disciplines such as physics, chemistry, or other fields of science may enhance cognitive development. When provided early in education, knowledge visualization may help support children's innate abilities and enhance the integrative training of their minds. Repeating some content by enhancing the difficulty level engraves information into their minds.

Knowledge visualizations help to clarify technical, engineering, and computing applications by managing data and information. Tools such as networking, data mining, cloud computing, web search, topology, and graph theory support knowledge management and visualization. Sharing knowledge may involve the use of artificial intelligence and machine learning software.

Art has often been used to stimulate imagination when conveying knowledge visually. Disney's Magic Kingdom and then EPCOT (2023) developed interactive entertainment based on motion, illusion, and different applications of technologies. An architect Celestino Soddu (www.soddu.it) used the 'DNA of the city' (including inhabitants' age, professions, leisure time, activities, number of children, groups, etc.) as guiding data for the computer to build structures matching each city and related interior and product design. You can also visit the Pokras Lampas® Calligrafuturism Artwork entitled *The Void of Dreamers*, where each round of calligraphy is based on the artist's memories and reflections about the culture, future, different generations, and letter shapes. Graphics with explanatory power shape our lives, behavior, safety, courtesy, and daily routines and speed our actions.

Using metaphors in data, information, and knowledge visualization may serve for integrative science-art instruction using computer graphics and programming. Instruction in knowledge visualization, along with computing technologies (especially for bioinspired solutions and nanoscience), should be introduced early (STEAM, 2023; STEM to STEAM: Integrated Studies STEM to STEAM Resources Toolkit, 2023). Industries, corporations, companies, and online networked communications need an introduction of these changes.

Project 3. Visualize your milieu

This project invites you to look at everyday objects surrounding a place you use for learning or your workplace and the content of your writing or reading. Present this place's core, nature, and spirit as a picture.

While working on something – writing notes, reading a journal, working on a project, homework, assignment, or other task – fix your gaze on what you are working on. If you are writing numbers, show them as metaphors: for example, sketch them as animals dancing or climbing on a graph. While reading a paper, sketch the objects you are reading about. Devise fantastic metaphors for non-tangible ideas or nonconcrete concepts described in your writing or reading. Your sketches will express your spontaneous reaction to these texts rather than illustrate them realistically. For example, Figure 5.3 shows some sketches of

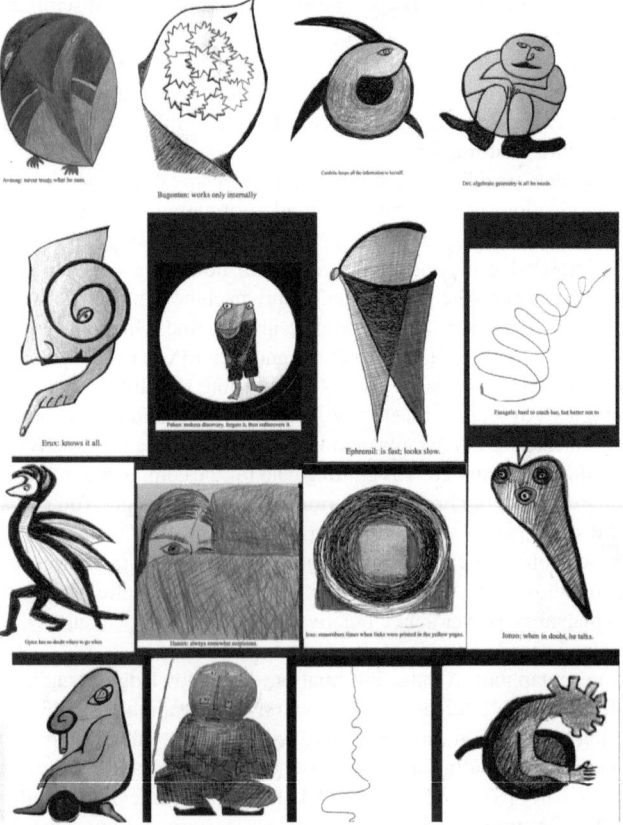

Figure 5.3 Sketches about objects taken from the surrounding environs

fictional creatures drawn as a response or a comment to the reading of a technical text, first about a concept of symmetry, then equations describing it, some practical solutions, and finally, a discussion of the results.

Arrange all your short sketches in a way that shows the relations between them. Apply color coding to emphasize these relations. For example, paint the visual metaphors you have drawn for numbers in the shades of one color, and assign other colors to other kinds of objects you choose to sketch. Look around yourself and then, in a few lines, draw a short sketch of your workplace. Draw it around all images you have already done as a background for the picture. Add some written content, either as captions explaining your decisions about shapes and colors or by writing a verse that would agree with the message created by your visualization and emphasize a psychological background of your decisions.

Our aesthetic emotions may depend on our physical state and feelings when we look at an object. Your final colorful composition may reflect your current mood, and after a few days, you will create quite different metaphors and shapes and choose another color palette.

Strategies for introducing novelty in instruction

Introduction

Projects about the visual way of learning support inclusive, integrated instruction through the arts conducted with gender neutrality and attentive to individual learners' strong intelligences (Gardner, 1993, 2000). They are focused on the visual power of description and explanation, for example, when designed as animations. The goals of projects offered by instructors should address the needs of learners related to their origins, natural abilities or skills, and cognitive potential. While designing learning projects, it is essential to apply information about the recipients and their intellectual capacity described in Chapter 2. Individual consulting or discussing the learners' problems, solutions, and successes may be delivered via Zoom, EonReality Coliseum, Skype, or similar systems.

Visual storytelling

Creating characters for visual storytelling

Visual storytelling, now interactive and immersive, is a tool of choice to communicate knowledge. In most cases, we introduce characters who actively transfer the content. Many kinds of AI services offer ready characters based on existing ones. However, it is much better to create new characters and make them suitable to convey novelty. Moreover, solutions offered by the AI services will improve when you add new prompts after completing the first part of the task. Creating a nasty, ugly, or repulsive character is not advisable because it will damage communication. The following text describes several projects.

Every day we encounter characters in various settings, including communication, learning, productive work in business, marketing, communication media, and artistic activities. We may assign many occupations to our characters, for example, an analyst, psychoanalyst, architect, logician, commander, diplomat, debater, advocate, mediator, defender, consul, explorer, virtuoso, entrepreneur, entertainer, or adventurer.

Characters may appear in a story in various forms.

- Avatars represent us or somebody else on the internet, in video games, in virtual reality, or in other places.
- Emoticons are made by keyboard characters such as :-) or :-(
- Emoji images are small images or icons in electronic messages showing emotions, such as smiley faces.
- Favicons are associated with the websites (that show a tab icon, URL icons, and bookmark icons).
- Logos are adopted by companies to identify organizations' products and more.
- Other visual signals.

Contrary to typographic emoticons, emoji are actual characters, often animated, and show facial expressions, everyday objects, animals, and more. Some time ago, *Wired* magazine invited stories made using six emojis. There are competitions for new effective emojis.

Creating characters supports communication, learning, visual storytelling, artistic activities, many types of dramatic art, productive work, and promotion of various kinds of business through marketing and advertising, among many other options. Including characters in a project may add another dimension when they wander across time and space. A character rarely exists without any action unless it acts as a symbol, such as the Statue of Liberty. A story shows its personality, integrity, and essential traits simply and concisely. It is important to eliminate unnecessary details and decide what is crucial for delivering the primary message.

People have created many kinds of characters, for example:

- Puppets represent human or animal figures and add life and expression to a story.
- Fingers or hands of an operator move hand puppets.
- Shadow puppets, as flat cutouts or 3D ones, are held against a screen and illuminated to cast a shadow.
- Rod puppets are controlled from below by long, thin rods.
- Marionettes are controlled from above by strings.

People make puppets from metal, ceramics, plastic, wood, socks, and 3D-print or knit them. Jim Henson developed characters for *The Muppet Show* and the *Sesame Street* television series. In the former Czechoslovakia (now the Czech Republic), there have been over 2,000 puppet theaters with artists such as Jan Svankmeyer and Brothers Quay contributing to spectacles.

The easiest way to design characters is by creating glove- and sock-puppet heads or paper-bag puppets. However, before deciding what characters you will create, review the existing ones, and reflect on why they are favorites. For example:

Curious George – curious
R2D2 – cute, productive, communicative
Superman – helpful, protective
Horse (chess) – clever
A dice – lucky
Apple – nutritional
Magic flute – beautiful
Winnie the Pooh – philosophical
Anatol – cheese connoisseur
Oscar the Grouch – grouchy
The Frog – a prince
Mickey Mouse – brave, playful
Bambi – horror
Frodo – brave
Felix the Cat – eventful
Mickey Mouse – cute, playful
Zorro on a horse – supportive
Willie E. Coyote – stubborn
Road Runner – smart
Robin Hood and Robyn Hood – brave, supportive

To create characters, it is helpful to sketch and draw daily as a way of thinking and to use a camera when accuracy is needed. As an interface between imparting messages visually and verbally, drawings and sketches of characters make effective tools for visual storytelling. Characters deliver emotions – visceral, emphatic, or voyeuristic – and evoke emotional responses from the audience.

In a digital, interactive, and shared environment, computing-based characters replace hand-drawn characters. This action often starts from the old cut-copy-and-paste operations, in many cases, without any concern about copyright issues. To add characters to visual storytelling, draw or sketch them on a computer, phone, watch, or tablet, work in traditional ways by sketching hand drawings with a pencil, a pen, or a brush, make 3D clay models, or 3D-print figures. Figure 5.4 presents a family of Kardelups.

Figure 5.4 A Kardelup family

The semiotic meaning conveyed by characters: using symbols, iconic objects, and metaphors

While drawing characters, you may include symbols that have the power to replace words and iconic objects representing other things or having a resemblance to them. Icons and symbols help to compress information visually. You may also apply a metaphor that indicates one thing as a representation of another. Metaphors make mental models and comparisons and show individual features of an imaginary character.

Characters may have different traits that make them fascinating, erudite, or romantic. They may be imaginary or real, literary or technology-based, magical or supernatural (Figure 5.5). Depending on the genre and purpose of the project, they may resemble humans, real or impossible creatures, or lifeless objects endowed with expression. However, remember that our audience may be tired of anthropomorphism (giving human traits, emotions, or intentions to nonhuman characters).

Characters in visual storytelling may serve for designing projects that support understanding science. Through their action, characters may visualize biological and physical events and their impact on people and their surroundings. They may explain techniques used in engineering and methods developed in new material science.

Telling stories visually

Visual storytelling, especially digital, may combine images, audio, and video clips with interactive, immersive, virtual, and artificial reality, so this is both visual and verbal communication. For example, a small board is connected to reality and obeys commands to display, collect data, etc.

Many would agree that whatever we do can become a story. It is a matter of finding a thought-provoking theme, transferring it into a specific platform, and

Figure 5.5 Threeall

telling it well, using proper tools, formats, characters, props, and backgrounds. Our actions depend on our audiences and agree with the tools used.

The storytelling projects may be finished, unending, or never-ending, looped animations or videos. They may communicate meaningful notions, introduce science-related themes such as cosmos or energy, and provide visual/verbal forms, e.g., beyond-the-text illustration, scientific illustration, literary illustration, and many more. We may apply storytelling tools to literary/visual genres such as:

- A sketch in its various stages, from doodles to rendered visual statements
- A novel, interactive novel, children's book
- Drama: a script for a theatrical play
- A script for filming, video, and moving images
- A music score set visually: a libretto for an opera, operetta, ballet, or musical
- Poetry: a poem, limerick, haiku, and more
- E-poetry and visual poetry, e-books, interactive poems
- Sequential art such as a graphic novel, comic book, visual narrative, animation, video, etc.
- Electronic novel, motion graphic narrative, storyboarding for digital media
- Social media–based environment, such as metaverses
- Anime for film and TV, Japanese style
- Manga for books and graphic novels, Japanese style
- Time-based visual material, animation, or a movie
- Interactive installation
- Arduino-based projects, Makey Makey, and Raspberry Pi physical computing
- Interpretive drawing going beyond basic images
- Technical writing and drawing
- Virtual reality in various ways: VR (virtual reality), AR (augmented reality), XR (extended reality: VR and AR), VRX world (a magazine dedicated to

animation, special effects, and computer-generated imagery published by the Animation World Network), winRAR (a demo program, file archiver utility for Windows)
- Machine learning with algorithms that learn by experience
- An immersive experience with 3D images surrounding a viewer

The stages of the work are called the preproduction, production, and postproduction of a story. The tasks include making a story with the roles of characters, defining a container for a story, building a storyline, making a storyboard followed by a visual presentation, using a video-editing program, writing captions, and finally, writing about a project.

DEVELOPING A STORY

Create a story, preferably telling new content rather than retelling old tales. Look around, think about your passions, what you know well, what happened today, and what you heard. First, present your theme, a subject you will deal with, and objects to which you will direct actions. Try to understand the underlying processes and forces defining the concept development. An inspiring character will tell a history of your fictional or research-based events.

The composition of the project starts with choosing a main character for a story along with its environment. Show features of an object or a set of characters and props. Show individually created environments, actions of invented characters or avatars, and changes in particular environments. Define the characters and factors such as dangers, characters' equipment, and implements, or symbiosis (known in biology as the interaction between different organisms living together; typically, this is a mutually beneficial relationship). Is your character big, fast, strong, and wise? Invent a conflict that may focus on apparent themes, such as a fish without water.

CONSTRUCTION OF A STORY

- Incorporate principal characters, such as a protagonist or a hero, in opposition to an antagonist.
- Define a setting.
- Build a conflict between a hero and a villain.
- Develop an action, suspense resulting from a threat and uncertainty, and then a surprise.
- Find a resolution.
- Make a conclusion and a moral.

Here, you can find a short example of working on a storyline:

- *An introduction of characters*: Paul and his three pets, Him, Wet, and Fur, were going up in an elevator to see a vet for their traumas.

- *A conflict:* The elevator got stuck, so the characters are stuck. The elevator stopped suddenly, making them all wonder.
- *A resolution, an action to solve the trouble, a rescue*: The vet saw their problem on the phone and ran to the rescue.
- *The consequences, treatment > no fear*: He treated their traumas right in the elevator, so they would not be afraid to return the next time.
- *A moral, a careful choice advised*: Select your vet carefully.

Examine how characters collaborate in the story. Typically, we develop a set of protagonists and antagonists and attribute the good and bad features of their characters, personalities, and beliefs to them. This action forces our audience members to take sides. They want or are told to wish one group of characters well. They want to convert the other group members to a proper set of beliefs or see them go and lose their way. However, does a lawyer defending someone who did something unacceptable by society feel bad about it? How one justifies it? A character who defends one's values or the purposes of his group can be seen by the audience as an evil character, or the audience can see these actions as bravery and goodwill. Do we need 'a good one' and 'a bad one' to tell a story? Do we need to think in black and white (such as yes/no, good/bad) and tell about our views and beliefs? Do we have to take sides and wish well or not so well? Perhaps games, sports, and gaming are the answers. In Figure 5.6, the heroes of a story may not be aware that somebody is overhearing their conversation. Figure 5.7 shows a moment of conflict or a threat. Make the whole project entertaining. Also, color adds emotion to time-based storytelling, as it does for visual arts. A narrative or animation may contain not only visuals or

Figure 5.6 The overheard conversation

Figure 5.7 A conflict between a hero and a villain

descriptions but also dialogues. In that case, write the characters' conversations, opinions, arguments, and declarations.

A STORYLINE

To build a storyline, develop an action starting from an introduction of characters, a conflict, and a resolution, leading to a conclusion and a moral with some expectations for the future or a message the audience should take away with them, such as: be an organ or tissue donor or recycle paper. Write your short story like you would tell it on a cell phone to a friend or you would tell this story to an eighth grader. A storyline, the narrative of a research-based story, should be short, engaging, without visual cues, and should not require showing things. Consider possible ways of redesigning this storyline for a particular audience.

A CONTAINER FOR A STORY

A container for a story allows us to introduce various formats and styles of displays and storytelling. It defines how we deliver the story. The oral way of sharing information was predominant before the emergence of writing in prehistoric and ancient times, and it is still an effective way of communicating without written texts, for example, in kindergartens. Old, popular tales are now treated this way in various media. An example of a container may be a complete book containing essays, short stories, or an academic journal containing articles. It can also be a database containing academic journals or a website containing web pages. We may apply storytelling to many formats, with the story retold depending on the media: a graphic, computer graphics, comics, manga, animation, theater performance, VR, a game, movie, radio show, video, motion picture, motion graphic, or any interactive or immersive medium (a 3Dimensional image generated by a computer which appears to surround the user). By choosing a story container, we define how we will deliver the story and assign the work to team members.

Storytelling through animation goes through time to enhance dramatic actions, show suspenseful obstacles, and build tension before resolving a conflict. Animations affect the emotions of the audience. This is why it is helpful to define the audience before undertaking the work on a project. Emotions are for a time-based storyteller what colors are for a painter. While you learn about storytelling-based animation, use a video-editing program. A 2D animation can be in a simple image format GIF (graphics interchange format) or an animation using the Motion workspace in Adobe Photoshop, where layers are employed. There are also many apps for doing this.

A particular setting may be appropriate for telling the same story for another medium, for a different type of delivery. For example, one may write a short story and then rework it in several ways, thus creating various types of containers for the same story. Choose specific settings and styles and then visualize the writing by assigning to the story some visual containers such as illustrations, video, or animation for this particular set. Characters and avatars may convey their stories through music, dance, and text. A computer program or computer graphics may define a container for a story.

A STORYBOARD

After the storyline is ready, create a storyboard – a graphic organizer of your storyline, a set of drawings or sketches going together with descriptions, dialogue, and directions. It shows the images you will make for your production. Storyboards are used for the previsualization of scenes or sequences before filming. The 2D or 3D storyboard software (Gackowski, 2015) and free printable templates are available online (e.g., Studiobinder, 2023; Canva, 2023). Animatic is a very short introductory version of a movie, made by shooting successive sections of a storyboard and adding a soundtrack. An animatic is often used as an initial version of a movie, video, animation, motion picture, motion graphic, or any interactive medium.

Finally, work on keyframes, find the resolution, draw conclusions, and make the whole project entertaining. Write captions and describe the project. Then it all goes to marketing and distribution.

Activity: Design a robot character with abilities

Sketch a robot and then assign 10 points to be spread between its abilities: speed, intelligence, and power. On a scale of 1 to 10, assign numbers for each of its three abilities. The total may not exceed 10. How much would you distribute for intelligence, and how much for speed and power? What would be a story about the one with higher intelligence and speed, which would win over the one with higher power? Alternatively, the one with better power and speed but with lower intelligence would take down the one with higher intelligence. You may add more abilities and skills, such as musical, math, and logic abilities and the skills of jumping, throwing, combat strength, and durability. However, the points you will assign to your robot should not exceed the total you set up at the beginning.

Think about the project's recipients, and choose materials, colors, and typography. Consider who the audience is. Every audience may have its own needs. Typography – setting types and the appearance of printed material – makes the written part of the project legible, readable, and appealing to the intended recipients. For example, small type is ill-suited for small children and older people, while high school students may consider a big font size offensive.

Activity: Interpretations

As an exercise, before devising what your characters are saying, set the TV on mute when people talk on the screen. If you have a group of people in the room, ask everybody to look at the TV screen (muted) and speak loudly what each of the performers is saying. If you are alone in the room, speak for the most active actor. Follow the types of emotions the performers or anchors are displaying. Speak intuitively, then write it from memory or have someone write it. It will be your own script for the story based on the conversation you witnessed on the screen but could not hear.

Pick one template that meets your aesthetic and intellectual needs. Select your keyframes, defining the starting and ending points for smooth transitions between images. A set of keyframes will define the movements seen by a viewer. Insert your major keyframes into the template. You may google free template examples and find online videos about this technique, some of them using the 3D printing technique, for example, a video about the 3D Rapid Storyboarding prototype (Gackowski, 2015).

Write captions under every change that will explain each dynamic action.

Learning can be accessed at no charge via some public libraries. It is a valuable collection of videos developed by professional instructors teaching various skills, for example, at Lynda.com, now LinkedIn (*www.linkedin.com/learning/?trk=lynda_redirect_learning*). Other companies are offering online videos to be used for learning. One can learn particular software skills, build a resume, or develop successful prompts for an AI.

Below are examples of animation apps you can use.

Procreate: *www.appconner.com/app-procreate?utm_source=bing&utm_medium=cpc&utm_campaign=225_afcapp_us_harry_1129_0.14&utm_term=Procreate%20app%20download&utm_content=2682_Procreate*

Adobe Character Animator: *https://helpx.adobe.com/adobe-character-animator/how-to/adobe-character-animator.html?SDID=KQOYR&mv=search&s_kwcid=AL!3085!10!79439847006424!79440121154465&ef_id=996477c9974811993 7781430928bc915:G:s&mv=search*

Express Animate Software: Motion Graphics and Animation Software for Mac: *www.nchsoftware.com/animation/index.html?*

Clip Studio Paint: The artist's app for drawing and painting: *www.clipstudio. net/en/?utm_source=bing&utm_medium=cpc&utm_campaign=bing_ad_text &msclkid=e1f3e5a45a72148362a2865d5fb18ad5*

FlipaClip: Create a 2D Animation (2023): *https://play.google.com/store/apps/details?id=com.vblast.flipaclip&gl=US*

Animation Desk® Draw & Animate 4+: Create animation with us: *Kdan Mobile Software LTD*, Designed for iPad. *https://apps.apple.com/us/app/animation-desk-draw-animate/id946346179*

DESCRIBING THE FINISHED PROJECT

Write a research-based paper on a visual story you created. It may describe two projects: developing characters and creating visual storytelling. It should be at least three pages long. Commit your storyboard to paper. Describe and sketch the main characters and the way they tell the story. Focus on how the style of sketches you created affects your writing technique and how the visual power of illustrations makes your text unique. Remember to write the parts typical of a research paper: an Abstract, Keywords, Conclusion, and a Moral explanation of the purpose of the story (for example, to entertain, promote something, make people give blood, fight against pollution, or stop smoking), followed by References. When you find some information online, include a URL informing readers where it can be found. Since websites are modified, write when you accessed it after providing a link. While illustrating your paper, place captions under all illustrations.

Serious games

In his book *Homo Ludens: A Study of the Play Element in Culture*, Dutch historian and cultural theorist Johan Huizinga (2016) said that play is the essence of culture. Learning can be strongly supported by playing (Lillemyr, 2020). Humans (both children and adults) and other mammals and birds are interested in playing. Play may be an activity on one's own, shaped as a role-playing game, or rule-based play governed by codes. It may include construction and experimentation or be physically related to body movements (Lillemyr et al., 2020).

Many instructors, teachers, and trainers consider serious games superb tools that stimulate, motivate, and entertain learners while supporting their studies. When an instructor or a teacher encounters uneasiness resulting from the learners' weak motivation, interest, or short attention span, serious games may serve as a way of dealing with this challenging situation. They may encourage co-workers or classmates to work together. Designing a game may help developers practice analytical thinking and support building skills needed to analyze data. When we write a story and design a game around it, we must look at its theme from various angles. We would think about various factors that change the action. Finally, solving the conflict in an unexpected way then drawing a conclusion and a moral may provide some fun.

A multitude of existing serious games helps learners in everything they explore, from math, programming, and science to art, languages, geography, and history. Instructors may engage individual learners and build a sense of teamwork in players.

An even better way of applying games as a tool for learning is to encourage individuals or team participants to design a serious game about a subject they are currently studying. Organizers of a Gamegala2023 competition encourage players, "Instead of just playing games, be a game creator and developer!" After designing a game on a selected theme, creating a real game requires the cooperation of team participants who are not afraid of coding. However, it is also possible to find a no-code game-development app online, e.g., App Master (2021).

As a part of the video game industry, serious games belong to a category of serious storytelling, which was first examined thoroughly by Artur Lugmayr et al. (2016). In serious storytelling, the narration reflects a thoughtful action beyond entertainment. Active learning facilitated by creating games and social interactions among participants may support traditional strategies: lectures delivered by teachers followed by rote learning and multiple-choice testing.

WORKING ON PROJECTS INVOLVING SERIOUS GAMES

Designing a new game involves the first task of writing a script. Generally, scripts are the texts written for an entertainment production – video games and also plays, films, radio broadcasts, and more. To make a video game script, first, write a plot, a narrative for a story, choose characters, and write dialogues that the leading players and nonplayer characters will tell. State the purpose, objectives, needs, and benefits of the game. Visual storytelling and creating characters were described before. After gaining practical skills in storytelling, creating characters, and making a storyline, you may choose some objects you are studying and use them to create characters for a game. Then it is time to create characters related to the topic learned, a storyline, and an informative background, and link it all with a subject of study.

All this work will enable you to originate a game, just making a general design telling how the game will look and what it will present, without going into the interaction with a user interface, input such as a joystick or keyboard, visual feedback as video or virtual reality, and the gaming techniques such as character generation, 3D modeling, scriptwriting, working on texture and lighting, etc. You may sketch characters or find many kinds of free 3D characters online (for example, free cartoon characters, 3D models, cgtrader, *www. cgtrader.com/free-3d-models/cartoon-character*, or free 3D character models, Blender Kit, *www.blenderkit.com/asset-gallery*). Many file formats include 3d Models Free New & Unrated Formats .obj (OBJ) .fbx (Autodesk FBX) .max (3DS Max) .c4d (Cinema 4D) .ma/mb (Maya) .blend (Blender) .unitypackage (Unity Game Engine) .upk/uasset (Unreal Game Engine). Available in many file formats including MAX, OBJ, FBX, 3DS, STL, C4D, BLEND, MA, and MB Find professional Cartoon character 3d Models for any 3d design projects like virtual reality (VR), augmented reality (AR), games, 3d visualization, or animation.

Project 4. Designing a chemistry-/physics-related game about atoms and particles

For this game, apply your knowledge about atoms and their particles and present the atom's particles as characters. We still need to study atoms and their role in building elements and chemical compounds even though they are no longer considered the smallest elements in nature.

An atom comprises several types of subatomic particles: a nucleus, electrons, and other short-lived particles. Some particles are attracted to the nucleus, while others avoid it. The nucleus consists of protons with a positive charge and neutrons, which are electrically neutral. Protons and neutrons are composed of particles called quarks. The number of protons defines the atomic number in the table of elements. Electrons have negative charges; they jump across orbitals – the wave patterns around the nucleus. They form an electron cloud. The nucleus is a small part of the atom. It was said that when a nucleus is compared to the atom, it would have a football stadium size compared to the whole earth. The number of protons is equal to the number of electrons, so the atom has a neutral charge.

Your game may be about adventures of atoms, and their parts would behave according to the rules defined by physics. We can see the relations between the atomic structures. The atom can be considered a container for the nucleus, electrons, and other particles. We can also see an atomic nucleus as a container for protons and neutrons.

For example, you may present an atom as a beehive and subatomic particles as honeybees with different roles as colony members. Inside a beehive is one queen, hundreds of drones (in spring and summer), and thousands of workers. The number of each kind of bee in the hive depends on the hive conditions and the time of year. Drone bees are males and are not present in a hive during winter. However, during spring and summer, there are usually several hundred drones. They are needed for mating with new queens (*https://carolinahoney-bees.com/bees-in-a-hive/*).

Characters representing different atoms and particles may have their adventures resulting from the laws of physics that rule the events under study. However, you can also go beyond these restrictions and pretend that your characters have arrived in another world. In that case, think about letting the players know the laws that rule the events.

Project 5. Designing a biology-related game about birds

Another game may be about several species of birds and a need to preserve them and their habitats. Choose from many biology- and environment-related issues that may serve to build a story for a game. The game may involve concepts such as bird migrations, a pecking order (a social organization within a flock of birds that determines which of them had the right to eat first), a life cycle of insects the birds are eating, and how it determines when they are happily satisfied to the full and when they are hungry. Action may also become dramatic due to the dangers of atmospheric events, drones, planes, big predatory birds, and other animals. Making a script and visuals for such a game may give you satisfaction because of the fulfillment of your motivation for supporting benevolent actions and the pleasure derived from your performance. For example, dramatize the unsolved conflict between wind farms and migratory birds and bats.

Wind farms are dangerous for birds and bats. Sources of energy including mining coal, wind farms, solar panels, and other projects are discussed regarding climate change and the need to protect the planet (Figure 5.8a). However, the birds are paradoxically endangered by current trends and actions aimed at preventing global climate change. A severe danger to birds and bats is wind farms that generate electricity from the wind's kinetic energy. Some white steel columns are over 400 feet tall and weigh over half a million pounds. Each column has three narrow blades over 150 feet long from the base to the tip of the highest blade, splayed at equal angles. Because of the heavy weight of steel

Figure 5.8a Electrical Syntax

columns, almost a thousand tons of concrete was poured down under them to bedrock, with vast amounts of carbon dioxide released into the atmosphere to produce this concrete. Blades convert wind into electricity unless the wind is too weak (below 27 miles/hour) or too strong.

For this reason, the average wind turbine produces only about 1/4 of its expected capacity (Kaufman, 2019). Collisions with wind turbines are killing birds and bats. For example, hundreds of birds of prey (red-tailed hawks, kestrels, and other species) died in Altamont, and about 75 to 110 golden eagles were killed each year (Kaufman, 2019). Golden eagles fly only during the day; the migrating songbirds are nocturnal and fly at night. In both cases, large numbers are killed by the whirling blades. No satisfying solution was found for this severe problem.

Solar panels would be safer for birds and bats (Figure 5.8b).

If you are interested in biology, technology, and protecting the environment, design a game about birds. Create a story first, for example, about birds, deciding to perform their migration above solar panels. Define a conflict in each story, and then find solutions against the wrongdoing characters' harmful actions. For example, migrating songbirds may share information about the low-risk pathways over the solar panels. They may feel safer showing the birds of prey a pathway over another solar panel region but not telling about their new pathway. Alternatively, focus on bats. A story will make the basis for the game.

Literary connotations. For those who are world literature savvy, fighting against the killing of birds and bats by the blades of wind turbines may remind them how Don Quixote mistook a windmill for a monstrous giant and attacked it with his lance. He is a main character of a Spanish literary classic, *The*

Figure 5.8b Peripheral vision. Migratory birds are safer with solar panels than with wind turbines

Ingenious Gentleman Don Quixote of La Mancha (1605, 1615), written by Miguel de Cervantes Saavedra (1547–1616). Many works of art created about Don Quixote can be seen online, for example, Gustave Doré: *Don Quijote de La Mancha and Sancho Panza* (1863), *https://commons.wikimedia.org/wiki/File:Don_Quijote_and_Sancho_Panza.jpg*, or a 1955 sketch by Pablo Picasso, *https://en.wikipedia.org/wiki/Don_Quixote_(Picasso)#/media/File:Don_Quixote_(1955)_by_Pablo_Picasso.jpg*.

Project 6. Designing a geology-related game about dynamic geology and deep-sea ecosystems

Imagine and sketch a character related to geological events in a deep ocean. For example, it may be an underwater volcano's eruption or an earthquake. To write a script for a game, identify yourself with such an avatar and its graphical representation on the screen. Imagine this character's various roles: it may fight or act peacefully to maintain the geological or ecological balance. You may design a character that impersonates a physical force or a factor that controls the weather. To enrich your figurative and descriptive imagery, search on the internet or use the help of AI to examine characters from Greek mythology. Also, old maps were decorated with impressive figures and showed personifications of the physical forces. Create a metaphor for an abstract concept such as temperature or pressure. It will act as a physical factor or a force. Gather information on what will happen when your avatar grows stronger and fights against disastrous meteorological events.

Choose the theme for your game. Actions of such characters may virtually control the course of events: preventing volcanic eruptions, mudslides, or meteorite impacts. After learning about the fault zones and tectonic plates in the ocean's deepest trenches, make a story going under a deep sea. While making a script, avoid resemblance to existing stories, such as *The Little Mermaid*.

For example, characters may be unmanned divers and robots which explore the dynamic events at the bottom of the ocean. They will fight against the troubles of deep-sea creatures resulting from human-made underwater explosions or tsunamis – enormous water waves resulting from earthquakes or underwater volcanic eruptions.

More about serious games, educational games, and gamification

A serious game is more than entertainment or competition. It is a video game that combines fun features with instruction, learning, information, and communication. The best serious games create situations in which one does not know that one is learning. Serious and educational games are similar, but a serious game is a product applied to enable motivated learning, while an educational game is a learning method. The use of games for educational purposes probably dates to Plato and Socrates. Based on the function of the games reviewed, Connolly et al. (2012) classified games as educational or serious: the purpose of educational games is more limited in scope, while serious games can be used for training and learning in many areas of business, trades, marketing, and government and nonprofit agencies (Connolly et al., 2012). Serious games became a discipline. Research studies have been focused on serious and educational games in terms of psychology, phenomenology, ethics, semiotics, ethnography, and law, among other approaches. They make a motivational instructional tool.

Serious games are custom-built from scratch for specific audiences, such as companies, individual public places, and private clients, and for specific purposes, e.g., surgeon or pilot training. They transfer knowledge and support abstract thinking, increase engagement, both intrinsic and extrinsic motivation and provide new skills. An intrinsically motivated game player can continue playing and improve their behavior and awareness.

Simulation games provide learning based on experience and observation. They support employees' training by offering state-of-the-art simulations of various situations and settings, which enables their analysis, prediction, or management. For students, they support learning various subjects, for example, biology, social studies, and business; they also provide entertainment. There are many free online simulation games for many platforms, such as airport management and sports management simulators, flight simulation for pilots, submarine environment, city, building, government simulators, and more. Many were produced decades ago (Wikipedia, 2023).

Educational games offer practice by using existing repurposed games. They serve students in a classroom with a teacher. They are attractive to learners, who can experiment and get instant feedback. Educational games enhance engagement, transfer knowledge, support abstract thinking, and help students learn new skills. However, the available assortment of different subjects could be more satisfying, and the instructors or teachers must be ready to guide game-based learning.

Gamification applies game elements in nongaming content, often on digital platforms. It often serves competitive people and is used in learning and training programs, public spaces, and classrooms. Gamification is more straightforward and less expensive, adds fun to tedious or boring tasks, intensifies competition within a company, and motivates employees (Laning, 2020).

There is evidence of a positive response to games from educators, who have stated that serious games could have positive effects on learning; transform the learning of students and other learners and improve knowledge retention (Denny, 2016, 2019); and enhance cognitive abilities that underlie learning: working memory and attentional capacity/executive function (Zheng & Gardner, 2017). Playing video games stimulates affective, cognitive, and communicational processes. Many games help the players retain information faster and longer; they enhance their psychomotor skills, reflexes, and reaction speed. Serious games stimulate curiosity and increase learners' motivation, when success in playing may improve self-esteem, give satisfaction, and a feeling of self-sufficiency. They provide gamified, interactive learning experiences and a feeling of adventure, thus adding variety to lectures and other instructional methods.

Serious games teach skills from typing to coding, which is considered by a growing number of authors a mandatory skill for today's youth. For example, the Gala competition, Gamegala 2023 (*https://gamegala.org/*), invites K–12 students to showcase digital game projects they have developed. Students may enter the competition individually or with a team. Students and other learners should know that serious games facilitate learning and are tools for study and assessment (Moro et al., 2020). Researchers have been studying the effectiveness, advantages, and challenges of using serious games to assess the effectiveness of using gaming for educational purposes and the impact of educational games on one's learning ability.

Players can discuss ideas and also immerse themselves in a knowledge-based game. They can learn together with their avatars and their 3D-scanned favorite objects. Lin et al. (2013) described a game-based remedial strategy: whenever, throughout the game, students were unable to answer a question, they received instruction for that question; students learned the answers and avoided the shortcomings of multiple-choice questions.

Research results confirmed that the Monopoly game enhanced the learning of mathematical concepts and was more effective than instructional videos because this game involved more senses. Apps on smartphones allow gamers to use many senses, such as those responding to temperature, pressure, and balance.

They use apps with touching ability and accelerometers (Connolly et al., 2012). VisionShiftStudios (2023) creates immersive experiences changing the perception of reality: e.g., a serious game, *Air Medic Sky 1: Interactive Patient Safety*, trains young doctors.

Serious games are used for medical simulations and by many industries, schools, corporations, and businesses; for example, there is a business category at the Gaia competition (GAIA, Games and Intelligent Animation at George Mason University, 2023). Serious games support scientific exploration, health care, defense, emergency management, engineering, and more. Works of art created through computer games or designed by a computer are called art games. Creating serious games aims to educate, train, or heal the players or advertise some issues or products. Along with pedagogical values, fun, and competition elements, serious games have some common features with simulation, especially in games related to flight, health science, or medical education and especially with a therapeutic perspective or a focus on supporting military resources.

There are many options when one wants to earn the game programming specialization. For example, University of Southern California offers a Master's of computer science with specialization in game development; Rocky Mountain College of Art + Design and The DeVry University (2023) College of Media Arts and Technology offers a multimedia design and development degree with a web game programming specialization. Southern New Hampshire University (2023) offers a game programming and development BS.

There are several institutions focused on serious games, for example, the Virginia Serious Game Institute, NYU Game Center (2023), Games and Intelligent Animation at George Mason University (GAIA, 2023), the Serious Games Society, and the Games and Learning Alliance (GALA, 2023). They organize national and international conferences. For example, the Serious Games Society (SGS) supports serious game research. This society, the Games and Learning Alliance, and local scientific teams annually organize the GALA Conference and competition. The Gala2022 special theme of the competition was Serious Games Which Support Literacy or Numeracy Skills (Gala, 2023). International Joint Conferences on Serious Games (JCSG) are organized annually, for example, the JCSG, 2023 on 26–27 October 2023, Trinity College Dublin, Ireland (JCSG, 2023), and the JCSG, 2022, Weimar, Germany (JCSG, 2022).

SOME NOTABLE SERIOUS GAMES

The top 10 serious games, recommended by Juliette Denny's *Growth Engineering* (2019), are listed here.

Flight Simulator, created in 1982 by Microsoft and still in use today, is a noncombat civil aviation simulator (*www.flightsimulator.com/40th anniversary*).

Minecraft (*www.minecraft.net/en-us*) is a well-known serious game, one of the first to show an explicit link between gaming and education. It is a Microsoft video game that helps children learn about many school subjects. Players build their 3D world using a series of different blocks. Regardless of players' skills, a game involves creativity and imagination. According to Shankland (2016), regular *Minecraft* playing helps players build brain cells and effectively introduces kids to computer programming. There is now an education edition explicitly crafted for schools.

Journey to the Wild Divine (*https://wilddivine.com/*) and the second game, *Journey to Wild Divine: Wisdom Quest* (*www.amazon.com/Wild-Divine-Biofeedback-Software-Hardware/dp/B00099YLPM*), address general health concerns. Interactive games with biofeedback sensors measure heart rate and skin conductance; players learn to control their breathing, build mental serenity, and reduce stress. Players can collect data and see real-time evidence of how the game impacted their well-being. Biofeedback helps people control their state of mind and serves as a learning stimulus (Corwin Bell, *www.visionshiftstudios.com*). In the second game, best-selling author Deepak Chopra, MD, guides users through brilliant 3D landscapes to reveal new breathing and meditation strategies.

Lightbot (*www.gameflare.com/online-game/light-bot/* and *apps.apple.com › us › app*

Lightbot Jr: Coding Puzzles for Ages 4+) supports learning STEAM skills. Children learn coding by solving puzzles. Two levels of the game are for ages 4–8 and 9 and up.

Re-Mission, a 2006 game, and *Re-Mission 2* (*www.gamesforchange.org/games/re-mission-2/*) help young cancer patients achieve better health and psychological outcomes while undergoing treatment. In each game, players find themselves inside a human body infected with cancer. They use various weapons and powers, such as chemo, radiation, and targeted cancer drugs, to fight and eradicate the disease.

Fate of the World (2011, *https://store.steampowered.com/app/80200/Fate_of_the_World/*) is about climate change. Players work with agents to stop global warming and energy shortages, cure food insecurity, and reverse nuclear weapons development.

Whyville (1999, *www.whyville.net/smmk/nice#what*) serves as a virtual world for children aged 8–15 and offers them more than 100 games, including the basics of chess, robot programming, and math puzzles. The games enhance creativity and develop critical thinking, decision-making, and entrepreneurship skills.

City One (IBM, 2010, *https://gamesforcities.com/database/cityone-a-smarter-planet-game/*) is a city-building simulation that addresses global city planning concerns. Civil engineers and city officials could use IBM *City One* to navigate real-world issues, like energy use, water consumption, and retail growth, and create solutions.

DragonBox (Elements, *https://dragonbox.com/products/elements*) is an online game that teaches children to master geometry. Players explore the properties of basic shapes and solve geometry problems. Versions of the game are designed for home use and educators.

Pulse!! (*www.gamedeveloper.com/pc/sgs-feature-i-pulse-i-first-person-healthcare-system-simulation-*) simulates aspects of surgery. Players move through learning modules and practice on virtual patients. It is safe and ultrarealistic, sold exclusively to the medical industry, and not a commercial release.

Pacific (*https://arc-institute.com/en/serious-business-games-2/pacific*) teaches management strategy with a survivalist spin. In the Pacific, the player and his friends crash their hot air balloon. The player must use leadership, management, and teamwork skills to repair the hot air balloon and get everyone home safely. The international game is sweeping corporations all around the world and is available in several languages, including English, Spanish, French, and German.

A selection of serious games is based on an article by Juliette Denny (2019). She declared war on dull online learning and stated, "Each of these games, along with countless others not included here, model the ways serious games can not only transform how we learn but also help us develop marketable skills, autonomy, and empathy for others" (Denny, 2016).

STEAM, STREAM, and STEM programs and the labor market

In visual teaching and learning with STEAM, STREAM, and STEM programs, students collaborate on interactive art-making and learn how areas/disciplines intertwine: art, technology, material sciences, math, physics, chemistry, biophotonics, and more. Teaching and learning visually should include techniques in data, information, and knowledge visualization; teaching about coding in computer languages; applying artificial reality solutions; instruction in serious gaming; the inclusion of virtual reality in the school environment; and instruction on social platforms for the global exchange of thought. Novel strategies make up essential parts of this integrative program. Options to choose from beyond traditional art forms are video art, web comics, virtual reality games, 3D printing, 3D casting, laser cutting and marking, and more. Many instructors and teachers stress the need to introduce universal languages such as Latin, music, and mathematics. Also, learners may be informed about using nature-inspired computing-based methods, for example, artificial neuronal networks, evolutionary algorithms, swarm intelligence, genetic engineering techniques, and bio-inspired hardware systems. Students should know and practice computer-based learning techniques such as learning networks, telecollaboration, and online learning. They should examine copyright-related issues and learn about preparation for job applications. In further text, the discussion relates to the ways learners proceed toward solving problems visually.

STEM education, introduced in 2001 by scientific administrators at the US National Science Foundation (NSF), is an approach to teaching that combines science, technology, engineering, and math, but it does not include the arts. Many schools introduce comprehensive curricular models, including curricula based on the STEAM program (science, technology, engineering, arts, and mathematics). The STEAM program integrates subjects into a cohesive learning model based on real-world applications. Students learn how areas/disciplines intertwine: art, technology, material sciences, math, physics, chemistry, biophotonics, and more.

The idea of including art in school curricula has a long history. In 2008, Georgette Yakman, a Ph.D. student at the Curriculum & Instruction Virginia Polytechnic and State University, originated a new integrative education model and presented it at the Pupils Attitudes Towards Technology 2008 Annual Proceedings in the Netherlands. According to Yakman, "The STEAM = Science & Technology interpreted through Engineering & the Arts, all based in Mathematical elements . . . The traditional academic subjects of science, technology, engineering, arts, and mathematics can be structured into a framework by which to plan integrative curricula" (Yakman, 2008). Later on, Yakman and Lee (2012) introduced the STEAM education model to Korea. In the 1990s, the NSF introduced the concept of STREAM (science, technology, reading, engineering, arts, and mathematics education) (Holoworld, The Basics of STREAM Education: A Comprehensive Guide, 2023). The STREAM approach to learning can be considered especially important in the current framework of the inclusion of storytelling, animation, video making, and the introduction of AI software as integral parts of school curricula.

There is a demand for STEM and STEAM graduates because there is a deficit of skilled professionals in US industry, and the deficit of middle- and lower-skilled technicians is more significant yet. For example, every optics engineer needs at least 10 optics technicians to introduce innovation to the market (Schlett, 2022). However, the supply of skilled technicians is decreasing. According to Josanne DeNatale, people from various professions are unaware that they may start working as optics technicians. They do not even know that this field exists. Community colleges in the US do not train enough technicians. There were more openings for optical sciences-related jobs at the recruitment event than students who attended this recruitment event (Schlett, 2022).

Computer-based learning

This theme could be seen as self-explanatory and obvious, but regrettably, many school and work surroundings avoid teaching and learning with computers too often. The many learning opportunities offered on the internet include various courses, computer applications, and learning software such as for making 2D and 3D animation and modeling. Education technology courses and training courses for employers and employees in different areas enable learners to control and evaluate their training. Software- or web-based training programs are offered offline, individualized, and as self-paced learning activities. Courses are usually standardized and consistent and often include tutorials, drill and practice exercises, problem-solving challenges, simulations, games, and assessment tools. For example, Adobe Learning Manager is a personalized learning tool for employee skilling. Microworld learning environments allow exploring a specific domain of knowledge.

In learning networks, telecollaboration, and online learning, students not only communicate and discuss ideas but also immerse themselves in knowledge-based games or learn together using their own avatars and -scanned favorite objects. Collaborative efforts bridge the gaps between various ways we draw, code, describe, or otherwise explore nature. Courses in educational technology are sometimes called computer studies or information and communication technology.

COPYRIGHT-RELATED ISSUES

One of the problems related to visual culture is that people have to understand and obey copyright-related laws. While in school, they are protected by educational settings and fair-use policies. However, when they graduate, all the work done with internet-based resources becomes a potential infringement in the eyes of prospective employers. Graduation is probably the busiest time in a student's life. Replacing copyrighted images with a private collection of visuals becomes almost impossible when one tries to graduate and apply for jobs.

Most employers are happy to see visual literacy applied in job applications. However, companies are quite afraid of copyright infringements because big companies, when detecting any sign of it in our networked, web-based environment, simply send a letter demanding typically $1,500, and if it is not paid on time, their lawyers follow up with a possible expensive lawsuit. For this reason, the learning process can be supported by collecting students' own visuals and applying them to their studies. Many students feel fear of drawing or sketching, and it is easier for them to come up with photographic materials on the subject. Using various devices, when guided in computer graphics, students create their own images related to their topics. They also learn from books and materials that aid in outputting their efforts. There are many copyright-related resources, but sometimes, it may be challenging to see when they are entirely free and under which conditions. It may be troublesome to prove one's copyright or secure permission.

Conclusion

The pace quickens in current developments in many domains. Every day brings news about advances in diagnosing and treating medical conditions, sustainable actions aimed at saving ecological balance, progress in computing, and the arrival of artificial intelligence and machine learning. Advancements in technologies based on optical sciences and knowledge about light make it indispensable for hiring candidates to have technological and visual literacy and familiarity with visual media and arts. New technologies involve a massive, still-growing workforce, which needs to understand visual ways of communication. Learning about techniques based on light and vision enhances visual communication.

Collaborative input related to scientific content may modify the format, goals, and content of instruction through the arts. The cooperation of mathematicians, computing scientists, and artists results in the introduction of visuals as a cognitive tool to visualize our knowledge about the world's structure. A need for integrative learning is evident because it supports preparation for success in finding a satisfying workplace for a new employee.

Novel strategies in instruction and learning visually combine information technologies with visual ways of conveying scientific content. Information about progress in science and information technologies, along with instruction in new art media, needs to be included in the school curriculum. Art is now incorporated into many disciplines, so instruction is necessary for employers, marketers (know your current audience), prompt writers, recruiters, employees' instructors, college students, high school students, and more. Developing one's skills in visual presentation allows one to be current in many nonart studies, research, and work.

6 Collaboration of professionals
Old and new examples

Introduction

As a rule, science is a collaborative endeavor. This chapter provides examples of visual shortcuts for grasping attention and comprehension of novelty. It tells how inventions, discoveries, and developments overlap across disciplines and how the teamwork of professionals shares areas of their interest and competence. Selected examples of collaborative research and visual presentation of knowledge may serve the instructors and learners as the themes for learning projects. Working on projects about collective achievements requires integrative learning focused on several fields of science at the same time. The application of visual and verbal communication devices may enhance collaborative projects. Devices may include film, performance, novels, graphic design, visualization, virtual reality, and more. With new factors such as the shortening of an attention span to 8 seconds, the greater the number of people making teamwork, the more seconds of focused work are offered, which becomes a valuable contribution.

The wisdom that came from living under the pandemic conditions brought a need for actions that require cooperative behavior, multitasking, outsourcing, the use of Zoom sessions, videoconferencing, and use of cloud phone systems, webinars, chats, AI apps, and other platforms. Multitasking refers to performing a couple of tasks simultaneously by a person or a computer. It demands concentrating in the middle of distractions while working at home or in the workplace. Many employers look for persons who can multitask successfully, for example, to handle emails, text messages, phone calls, and Zoom sessions while making in-person contacts with colleagues and clients and outsourcing to third-party firms, domestic or foreign, to perform tasks or services making some specialized parts of a project or a production. Marketing, IT services, website development, recruitment, manufacturing, and distribution are often objects of outsourcing. Due to the collaboration between specialists in optics, photonics, microscopy, computing, electronics, and microbiology, researchers can find symmetrical structures at the nano level (Putz, 2022). In human culture, aesthetics, and architecture, collective efforts create and preserve our surroundings. Gaming also demands teamwork, and so it is changing storytelling.

DOI: 10.4324/9781032705347-6

A serious gaming discipline could develop by collaboration among computing specialists, coders, educators, and scientists.

We are still learning from war experiences. The resilience of the defenders of their countries taught us about the importance of support, collaboration, and data sharing on an international scale and with tightened security. Everyday videos evoke widespread supportive responses and evidence of the importance of visual communication. Many significant projects that enhanced knowledge and improved human lives were possible due to the collaboration of people with various specializations. The examples discussed in what follows examine the interdisciplinary efforts of the cooperating specialists who made inventions and developed well-known devices or applications in different places and periods.

Current insight about prehistoric ways of imaging the cosmos

Present-day research conducted by archeologists teaches us about visual documentation made by prehistoric inhabitants of caves. Many professionals, such as physicists, chemists, computing and computer graphics specialists, statisticians, linguists, and historians, support the findings of archeologists. Before inventing writing systems in prehistoric times, people communicated, shared, and documented their knowledge through images. In ancient times, people communicated in both visual and verbal ways. Art served to visualize and transfer knowledge about the cosmos in all periods and showed the meanings and rules behind celestial events. Current data about cosmic events confirm observations made by prehistoric people.

Our understanding of prehistoric and ancient times has changed due to the findings of many professionals. The cooperation of technical specialists enabled developing and applying several technologies, instruments, methods, and analyses. Present-day methods used by researchers include radiocarbon dating data (Valladas et al., 1992, 2013, 2017; Quiles et al., 2016; Sweatman & Coombs, 2020; Sweatman, 2020; and more), statistical tests of probability checking if the findings were accidental (Sweatman, 2020), chemical dating of the paints used to show the positions of stars in ancient times, confirming those data by software (Sweatman, 2018), and other computerized methods.

Analyses of cave paintings performed in the 21st century revolutionized our understanding of prehistoric populations' perceptions. The prehistoric processes, such as the origins of civilization in West Eurasia and the Paleolithic-to-Neolithic transition, are now seen differently. Previously, authors linked the beginnings of civilization with the rise of agriculture. Radiocarbon data suggest that massive megalithic constructions with advanced artistry and symbolism preceded agriculture. Marshack (1972) wrote that prehistoric people explained how the world works through stories, images, and symbols. These artistic expressions were the early forms of scientific understanding. Knowledge and understanding displayed by prehistoric cave dwellers tell about their intellectual power as hardly any different from us today.

Prehistoric cave dwellers left more than handprints and depictions of animals in some of the oldest caves across Europe, artifacts that have been dated to about 40,000 years ago. Animal symbols running around on cave ceilings were previously seen as images of animal spirits or records of successful hunting. However, they may have represented the cosmos (Marchant, 2020). They pictured stars' constellations and represented the precession of the equinoxes, telling about the path of the sun in the celestial sphere. Prehistoric people presented changes in planets' positions visually, telling about their understanding of shifts of the Earth's rotational axis and their effects. These shifts are now called the precession of the equinoxes. Each animal symbol represents a constellation corresponding to one of the solstices or equinoxes.

Astronomical insights, much more significant than archeologists previously believed, strongly predate writings of ancient Greeks and artifacts found in archeological sites on the continent of South America. The findings support a theory of multiple comets' impacts throughout human development. Also, advanced knowledge about stars' and planets' positions may have aided navigation on the open seas, thus supporting prehistoric human migration.

People used their knowledge about the positions of stars to keep track of time. Contemporary archeologists from around the world decoded animal symbolism of date-keeping.

Below are some examples of artistic imaging created in Paleolithic caves, with radiocarbon dating:

- Hohlenstein-Stadel cave, southern Germany, circa 38,000 BC
- Chauvet, northern Spain, 35,000 to 30,000 BC
- Lascaux, southern France, circa 17,000 BC
- Pech Merle, southern France, both 25,000 and 15,000 BC
- Altamira, northern Spain, circa 15,000 BC
- Göbekli Tepe, southern Turkey, circa 9500 to 8000 BC Çatalhöyük or
- *Çcatalöhuyuk*, in southern Anatolia Turkey, 7500 BC to 6400 BC

This knowledge is extremely ancient, and it was widespread. All sites used the same astronomy-based method of date-keeping, even though the art was separated in time by tens of thousands of years. Sweatman (2020) and Sweatman and Coombs (2020) state that the same constellations are used today in Europe. In particular, Pillar 43 at Gobekli Tepe (now Turkey) is essential. It is like a prehistoric Rosetta Stone because it allows the decoding of animal symbols. The Paleolithic Shaft Scene at Lascaux has a meaning similar to Neolithic Pillar 43 at Göbekli Tepe and Çatalhöyük. They display the same method for recording dates based on the precession of the equinoxes, with animal symbols representing an ancient zodiac.

Prehistoric artwork recorded events such as comets' destructive strikes on Earth by showing various configurations of animal symbols. Visual documentation of such a strike around 11,000 BC was found in stone carvings at a site called Gobekli Tepe. In the Lascaux Cave in France, a drawing featuring a dying man and several animals may record a comet strike around 15,200 BC. Similar findings in Spain and Germany show Paleolithic and Neolithic art featuring animal symbols.

Activity: Communication in an ancient way

Make a visual message using tools that were available to ancient people. Draw this message on a stone, brick, wood, or ceramic surface. Your visual storytelling may be about a recent adventure, an invitation to a party, an account of your troubles caused by the latest storm, notes about your visit to an amusement park or a zoo, notes about your learning new software, or whatever you choose to draw. Show your sketches to a friend and ask this person to translate them into words without your prompts. Then, do it again to make it more readable. You may want to do it with AI.

A mathematician, anthropologist, and architect work on architectural friezes

Facades of hundreds of old buildings in the village of Pirgi, a small town on the Greek island of Chios in the eastern Aegean Sea, have been all covered by the gray and white friezes. The façades of each house had different friezes and contained circles, squares, triangles, and rhomboids. Unique façades displayed a lively geometry and gave each house its aesthetic value and distinctive identity.

A mathematician, an anthropologist, and an architect examined over 400 friezes (James et al., 2004) to explain this powerful art form and to discover the mathematical structure underlying the friezes. The authors applied a mathematical group theory and a computer analysis to the structure and organization of these old friezes in Pirgi. They discovered that the frieze artists obeyed color-reversing rules of the friezes' mathematical structure. The symmetry of color-reversing friezes revealed five basic actions of moving an image: translation, vertical and horizontal mirror, half-turn, and glide reflection.

Activity: Sketching a frieze and a meander

As a brief distraction from other tasks, make quick sketches of a **frieze**, a horizontal stretch of figures (simple sketches of people, animals, plants, or inanimate objects). You can save them for future reference to decorate something you want to design later.

Sketch also a **meander** motif for a decorative border, a pattern of winding or interlocking lines. Draw a continuous line that makes a repeated motif. It can be a motif with straight lines with right angles or rounded shapes that may be seen as water waves.

After that, define prompts for an AI-related application, look at the machine's solutions, and follow up with new prompts to arrive at a solution that would satisfy you. Then, look at examples provided online.

Collaboration of Albert Einstein, Marcel Grossmann, and Michele Besso

When Albert Einstein worked on his generalized and then special theory of relativity and the theory of gravitation, he collaborated with his friends from Zurich University: mathematician Marcel Grossmann and engineer Michele Besso; they helped him to create a mathematical background (Isaacson, 2008). Over one hundred years ago, Albert Einstein received the 1921 Nobel Prize in Physics. His theories influenced the philosophy of science and modern physics. He introduced the mass-energy equivalence $E = mc^2$, discovered the law of the photoelectric effect, contributed to the development of quantum mechanics theory, and explained Brownian motion (random motion of particles suspended in a liquid or a gas). Collaboration with his friends supported Albert Einstein in developing the theoretical basis for his physical theories, since he always preferred exploring the rules and mysteries of physics over studying advanced mathematical problems.

Cooperation of Lady Ada Lovelace and Charles Babbage on early concepts about developing a computer

A mathematician, Lady Ada Lovelace (1815–1852), and a mathematician and inventor, Charles Babbage (1791–1871), cooperated from 1833 on for a long time when each of them conceived different concepts and elements of calculating machines, which were the forerunners of the current computers.

Ada Lovelace discovered the possibility of programming a computer so it would perform complex calculations. She described an analytical engine and wrote an example of a sequence of mathematical operations, how to calculate the Bernoulli numbers. Computer historians consider it the first computer program (Gregersen, 2023). The programming language Ada was named to honor Ada Lovelace.

Charles Babbage had designed a calculating digital machine called the Difference Engine which operated on discrete (consisting of specific sets of values) decimal digits 0 through 9. He also designed plans for an automatic digital computer named the Analytical Engine, which would perform any arithmetic operation controlled by the punched cards storing numbers. Babbage planned that his machine would be steam driven (Britannica, 2023, Charles Babbage).

About 150 years later, Steve Jobs (1955–2011) and Steve Wozniak (b. 1950), who co-founded Apple Computer company in 1976, developed the Macintosh in 1984, the first mass-produced computer with a graphical user interface.

James Watson and Francis Crick discovering a double helix of the DNA molecule structure

In 1953, molecular biologist, geneticist, and zoologist James Watson, together with biophysicist, neuroscientist, and molecular biologist Francis Crick, discovered a nucleic acid double helix arrangement of the deoxyribonucleic (DNA) molecule. This structure has two chains linked by hydrogen bonds between pairs of nucleotide bases: adenine (A) with thymine (T) and guanine (G) with cytosine (C). DNA carries genetic code information and generates protein synthesis by making copies of itself.

Many other researchers contributed with their expertise in other fields of knowledge. The organic chemist Alexander Todd defined the DNA backbone of the DNA molecule. A physicist and biophysicist, Maurice Wilkins, introduced the idea of studying DNA via X-ray crystallographic techniques. An X-ray crystallographer and chemist, Rosalind Franklin, obtained evidence of the DNA helical structure by making X-ray images of DNA fibers. Her X-ray diffraction images provided Watson with proof of the DNA structure. In 1962, Watson, Crick, and Wilkins were awarded the Nobel Prize in Physiology or Medicine; Rosalind Franklin was not awarded because, in 1958, she died at age 37 (Science History Institute, 2022).

Later on, at an advanced age, Watson had controversial views about the applications of his discoveries to genetic treatments for cancer, the relation between race and IQ, and the importance of genetic differences that favor intelligence in some populations, which cut his ties with previous collaborators and the contributors to the Human Genome Project.

The Human Genome Project and the Human Brain Project

The international research project called the Human Genome Project (2023), the world's largest biological project, was conducted in the United States, Canada, New Zealand, and Great Britain from 1990 to 2003 to identify, map, and sequence a complete set of human genes. The order and sequence of the nucleotide bases in DNA make a blueprint for an organism and provide instructions for creating its genetic code. The complete human genome contains 3 billion nucleotide base pairs in the correct order. Genes are sections of DNA that affect particular characteristics or conditions when they control the production of proteins. All human genomes are identical in 99.9% of genes.

In 1990, scientists worked first on sequencing about 20,500 genes present in human beings. Then in 2009, Radoje Drmanac (2009), with 64 coauthors, published an online *Science* magazine report about a DNA sequence completion of the human genome sequencing. The authors devised the efficient imaging of a genome sequencing platform. They investigated patterned nanoarrays of self-assembling DNA nanoballs. They sequenced three human genomes and identified 3.2 million to 4.5 million sequence variants per genome. According to the authors, this platform promises the understanding, treatment, and prevention of human diseases, among other applications.

Over 750 scientists from more than 20 countries working on the Human Brain Project (2023) contributed to the developments in research on neuroscience, computing, and brain-related medicine. Research on the brain's structure and function takes the form of an open-source modeling platform named the Virtual Brain. Super-resolution imaging techniques allow seeing in 3D the activity of brain molecules and investigating Alzheimer's and Parkinson's diseases.

Jennifer Doudna and Emmanuelle Charpentier performing a CRISPR gene-editing technique

DNA-editing technology was a revolutionary genetic engineering development. Jennifer Doudna and Emmanuelle Charpentier were awarded the 2020 Nobel Prize in Chemistry for their work in CRISPR gene editing (a molecular biology technique used for modifying living organisms' genomes). Specialists in many disciplines cooperated to decipher the gene-editing codes (Isaacson, 2021). Doudna cooperated with specialists in the biotechnology of genetic engineering who were taking a fragment of a virus, splicing it into the DNA of another virus, and just making artificial genes. This technique allowed scientists to work on gene therapy by putting the engineered DNA into human cells. In the 1990s and the 2000s, cooperation focused on gene editing by delivering the missing genes into the T-cells of patients using a CRISPR-Cas9 cutting enzyme. This step led to the founding of several companies working on CRISPR-Cas9, as well as lucrative patents.

The developments were achieved by collaboration and competition of the Doudna–Charpentier team with scientists from MIT, Stanford, and Harvard: George Church and Feng Zhang from the Harvard Medical School, Luciano Marraffini, and Virginijus Šikšnys, who focused on mechanistic studies of CRISPR-Cas, and many other teams of researchers who became the heroes of CRISPR that deserved attention. Eric Lander (2016) celebrated the heroes of CRISPR-Cas9 gene-editing tool when he wrote, "Scientific breakthroughs are rarely eureka moments, they are typically ensemble acts, played out over a decade or more, in which the cast becomes part of something greater than what any one of them could do alone." Therapies achieved included an infusion of the edited stem cells to cure sickle-cell disease, cancer, blindness, leukemia, Huntington's disease, and more. Computer programmers and biohackers contributed with their controversial inventions in biotechnology.

Recently, the CRISPR-Cas9 genome-editing technique allowed the deactivation of the genes for two pigmentation enzymes in the hummingbird bobtail squid (cephalopod) *Euprymna berryi*. This way, researchers from the Marine Biological Laboratory (MBL, MA) obtained a genetically engineered, nearly transparent albino squid. They could study *in vivo* the squid's neural activity by inserting a fluorescent dye into its optic lobe (Ahuja et al., 2023).

Activity: Making future collaborative projects

Think carefully about those of your actions that evoke your interest. Imagine things or issues that arouse your emotions. Let yourself pretend that you have been awarded a substantial grant that allows you to do what you want and work on a matter of your interest, for example, in health sciences, archeology, or actions to improve the climate. Consider what kinds of pursuits you would choose and what practitioners you would need to invite to cooperate. Write several sentences about your thoughts and illustrate them with short sketches. Try to make some constructive plans, not just limit yourself to the protest art or actions.

Art restoration as a cooperative action example

Conservation of art and cultural heritage combines disciplines and fields of knowledge. To reconstruct a precious artifact, people ask art historians for help. They look first for similarities of artifacts created in that time frame. The love for beauty and the search for messages, contexts, materials, techniques, and trades are based not only on art-historical analyses but on deep knowledge of physics, chemistry, photonics, history, and more. Researchers use fractal technologies and many photonics techniques to check if the artwork is genuine. Computational microscopy imaging and hyperspectral imaging analyze a wide range of light across the electromagnetic spectrum, not only primary colors; these imaging techniques are also used for defense, agriculture, and biomedical imaging (Freebody, 2022). Specialists examine the paints' composition and the way there were made. Radioactive dating methods include carbon dating and lead dating. Some chemical elements or compounds, for example, titanium, might have been unknown when the painting was created. For example, Rembrandt's masterpiece *The Night Watch,* painted in 1642, is now under the Rijksmuseum's multidisciplinary research and conservation project named Operation Night Watch, made in cooperation with chemistry and physics professors, painters, art historians, and other specialists. Recently, researchers used the X-ray powder diffraction mapping method and found in the painting rare traces of compounds called lead(II) formate ($Pb(HCOO)2$ and lead(II) formate hydroxide $Pb(HCOO)(OH)$, never reported in historical oil paints. This is the first report of the presence of lead formate in a historical painting (Quellette, 2023; Gonzalez et al., 2023). Researchers explore possible reasons for the formation of this compound in historical oil paint using microanalysis, synchrotron radiation, and infrared microscopy.

The task of checking the genuineness of the artwork is challenging. For example, in the 20th century, the Dutch forger Han van Meegeren made paintings resembling those of Johannes Vermeer (1632–1675) and other artists. He called them the 'just-discovered unknown works' of these artists. He made his paintings on canvases from the old times and applied similar chemicals as the old masters did, selling his works for millions until they were recognized as frauds. The radioactive lead dating method revealed slight differences in the isotope composition of lead. Additional trace elements differed in the 17th century from those in contemporary lead pigments. Investigations using gas chromatography confirmed the forgery.

Restoration of works of art begins by studying the artwork in the framework of historical and contemporary functions. The focus is on light. We cannot see the sky as it looked during Rembrandt's times. We do not have the data because meteorological technology was not invented yet. We can read written works and memoirs, conclude from observations, and design a map of factors present at a particular time. Nevertheless, whatever the art restorers do has to be nonpermanent and reversible, because the solutions they apply may need replacing according to technological progress. Thus, art conservation is one of the most collaborative tasks, demanding multidisciplinary studies, and conclusions should be drawn based on the love for art.

Activity: Imaginary painting

Imagine that you inherited a beautiful old painting. What would you do to evaluate the artwork? How would you preserve its condition and not damage it with unprofessional conservation attempts? Whom would you ask for help in this case? Sketch a picture of such imaginary heritage artwork that would make you happy. Would you prefer to inherit a portrait of one of your ancestors, a landscape showing where your family lived before, or a modernist painting telling that your great-grandfather was a recognized artist?

Working together to conquer the global pandemic

The cooperation of scientists grew in importance when the SARS coronavirus in 2003, and then the severe acute respiratory syndrome coronavirus 2 (SARS-CoV-2), rapidly spread among humans. DNA (deoxyribonucleic acid) and RNA (ribonucleic acid) are the molecules of the nucleic acids carrying biological information in cells, such as coding, decoding, regulation, and expression of genes. Many viruses have their genetic material in their DNA. In coronaviruses, the genetic material is RNA.

Specialists concentrated on genetics, biochemistry, pharmacology, and chemistry, along with biotech, biomedical, and CRISPR specialists, built a network of cross-disciplinary teamwork groups. Starting with the Innovative Genomics Institute (IGI), many hospital and academic labs, the Mayo Clinic team, Stanford, MIT, and Harvard developed testing facilities. Anthony Fauci, chief of infectious diseases at the National Institutes of Health, became the leading force in fighting the pandemic. However, the federal government administration refused to perform widespread testing of people.

The volunteer army cooperated with the specialists in developing coronavirus tests, then at-home tests, and then vaccines, both conventional ones and those employing nucleic acids. At first, they worked on producing the genetic vaccines after Chinese researchers published in 2020 the genetic sequence of the new coronavirus. DNA vaccines deliver the genetic code for the DNA or the messenger RNA to stimulate the immune system. Then, the RNA vaccines served as messenger RNA that instructed cells to make part of the spike protein on the surface of the coronavirus (Isaacson, 2021). Vaccines were developed in several places, and biohackers from Ukraine, California, and Mississippi used genetic material experimentally without regard to accepted standards and tested them on themselves. The CRISPR-based systems started to protect people without activating the immune system. Moreover, they could be easy to reprogram if a new virus would emerge.

Now, as Walter Isaacson put it (2021), we need not only scientists but humanists to proceed when the great pandemic has partially and temporarily receded: "We can feel our way together, step by step, preferably hand in hand."

Activity: Be aware

Devise a new sign, a song, or a gadget that would encourage your friends to take care of the still-existing danger of catching a contagious disease while traveling far away. Tired after a long-lasting pandemic, many people are no longer careful and often do not obey warnings. Make simple sketches of your solutions. Avoid signs and symbols that are already present. Add a captivating caption, then look at others' reactions to your messages. For example, you may want to make a little piece of wooden or wired jewelry so your friend can wear it when going on vacation.

Monitoring migratory birds

Millions of birds travel annually from northern regions such as Alaska, Canada, or Siberia to Southern America and then back. People, especially millions of birdwatchers, are concerned about the safety of migrating birds, with many species already endangered. Birdwatchers cannot see the flocks of traveling birds because most travel at night. They can observe how the songbirds drop to the ground at dusk to find a hiding place for a day or two to rebuild their strength.

Monitoring migratory birds and gathering knowledge necessary to protect birds requires the cooperation of people with many areas of specialization. Biologists have deep theories, run experiments to confirm them, and capture single birds to tag them with geolocating devices. Satellite tagging uses radio transmitters too heavy to be carried by most songbirds (Kaufman, 2019). It is a challenge for scientists and engineers to devise semiconductors and transmitters so small that they can be used to track miniature birds. Further collaboration on new technologies resulted in the development of a Motus (2023) system ('motus,' in Latin, means motion).

Motus is an international research network that uses a radio telemetry array to track the movement and behavior of small birds and other small animals (Motus: Wildlife Tagging System, 2023). Motus uses nanotags and digitally encoded radio transmitters. Nanotags carried by songbirds of many species enable researchers to investigate migratory movements of the smallest birds and many species of bats and large insects, such as dragonflies, through the Motus network (Lotek, 2023).

The nanotags broadcast selected frequencies several times each minute. These signals are received 24/7, year-around, on a stationary receiver mounted on a short tower. Signals can be detected within 9 miles. Five hundred fifty towers were deployed from 2012–2019 between Canada and northern South America. They detected individual songbirds over 250 million times, according to Kaufman (2019), who wrote, "The champion thrush evidently had flown without a pause across the Caribbean and across the eastern part of the United States, going from Colombia to Canada at an average speed of about forty-seven miles per hour."

Activity: Knowledgeable actions

Draw simple sketches showing how you would design squirrel-safe feeders and platforms for birds depending on their favorite diet: small grains of millet, large sunflower seeds, sweet nectar beloved by hummingbirds, and fatty food for chickadees, titmice, or marsh tits. Squirrels can jump several feet up and down and are natural acrobats, so your task takes work. You may also want to sketch the birds.

If you advocate for squirrels as pets and visitors to your house, design a squirrel feeder to be placed in a space that is allowed for them. Research first what you will put there, because some food known as being suitable for them might not be good, for example, acorns, elderberry, corn, pecan nuts, and ivy. Pine nuts are harmful to squirrels, while sweets, cakes, pizza, hot dogs, tacos, peanuts, cat and dog food, and junk food are unhealthy; the seeds or the skins of avocados are unhealthy because they contain persin, a natural toxin that contains cardiac glycosides which can cause digestive issues, respiratory difficulty, and heart problems.

Studying animal cognition

In her study on animal intelligence, specifically the cognitive abilities of African grey parrots, researcher Irene M. Pepperberg (1999) engaged several specialists in cooperation to widen the spectrum of the research themes. In the Massachusetts MIT Media Lab in Cambridge, some researchers inspired themselves with grey parrots' intelligence while studying computers as learning systems and working on machine learning. Others worked on constructing electronic bird sitters for parrots who suffer loneliness while their owners work. Parrots are very sociable birds, so they become psychotic, screech, and pluck their feathers when left alone for a whole day or longer.

People who study linguistics cooperated on a study about understanding how phonemes – individual sounds – produce a complete word. Young children and people with Down syndrome need to learn how to make new words by combining sounds. Grey parrots can do that, so cooperative studies on these themes play a part in advances in medicine and health sciences. The producer and actor Alan Alda documented the Pepperberg's parrots' skills in a PBS show and a video for the Scientific American Frontiers. Irene Pepperberg co-worked with psychologists while studying number concepts in children, such as counting and addition abilities, and tested math skills in grey parrots. Pepperberg (2017) analyzed how grey parrots see the world and how they perceive optical illusions and examined their vocal learning, symbolic communication, and social learning. With researchers specializing in evolutionary anthropology and apes' cognitive skills, she compared abstraction and cognitive processing abilities in grey parrots and apes.

Studies on animal cognition resulted in a discussion of the accepted theories about the origin of human cognitive abilities. Collaboration on comparative studies of cognitive development in humans and birds demonstrated that, as Irene Pepperberg stated, avian brains, with a size of a shelled walnut, are "equal and sometimes surpass humans concerning various cognitive tasks" (Pepperberg, 2008, p. 204).

Cooperation in science and medicine

Bioimaging, a field that explores biological structures and functions, involves the collaboration of specialists who draw information from sources such as light waves, nuclear magnetic resonance, X-rays, or ultrasounds. Researchers then create information visualizations in two, three, or four dimensions.

Looking at minute structures inside a human cell. The total internal reflection fluorescence system (Ross et al., 2022) serves for imaging, just below the surface, the minute structures, actions, and processes such as adhesion, hormone binding, neurotransmitter secretion, and membrane dynamics. Showing these themes requires the cooperation of biologists, physicians, biochemists, physicists, specialists in optics and photonics, computing experts, and technicians skilled in microscale engineering.

Developing a prosthetic limb requires the collaboration of surgeons with engineers who design and build the internal frame or skeleton of the prosthetic limb and a pylon that provides structural support. Physiologists and physicians decode neuromuscular signals that would control prosthetic limbs. Then the computing experts can build computer models to mimic human motions for a prosthetic arm, wrist, or hand. Instead of teaching a prosthetic limb the typical patterns of behavior and how to translate these actions into commands, artificial intelligence, and its part, machine learning, are put into action. The incoming signals train the machine's model of a prosthetic limb. The machine itself finds, decodes, and learns patterns in neuromuscular signals that control prosthetic limbs. Then, a machine statistically determines boundaries and makes decisions about actions. Computer programs follow the body signals that create movements, including communication between brain activity and body parts, which secures fluid motions. Programs improve themselves based on experience, not by an action of a programmer who wants to write a better algorithm to rule the motion. A person receives sensory input back from the prosthesis to electrodes implanted in the muscles and nerves of the amputated limb so that this person can feel the sensation of touch and control movements. Due to collaboration with specialists in 3D printing, robotic arms now incorporate 3D printed materials, which can reduce the price by up to 90% (Kosowatz, 2020).

Collaboration on laser-assisted underwater welding

Scientists in the Laser Zentrum in Hannover, Germany, collaborate with industry specialists and manufacturers of sensors, microsensors, measuring instruments, and submersible probe systems. Together, they develop laser-assisted underwater welding technology to replace the manual welding of electrodes (Industry News, Photonics Spectra, 2022). In laser beam–arc hybrid welding, the laser beam inserts energy to improve arc ignition. This technique will serve in projects requiring technical underwater constructions such as wind farms, coastal protection structures, and harbors. It also serves to dismantle the reactor vessels by applying underwater laser cutting.

Further cooperation will enable combining the underwater welding torch prototype with integrated laser and laser welding of 3D-printed components in small- and medium-sized businesses. In laser welding, the molding goes through the injection of plastic components. In the 3D printing additive process, thin strands of molten plastic are superimposed layer by layer. The adaptive system of regulated laser power would enable a uniform weld seam, even if the component itself is not uniform (Researchers Adapt Laser Welding for 3D-printed Parts, 2021).

Closing conclusions

One may conclude that several recommendations may ensue from this book's content about urgent issues to be put forward and solved:

- Need to apply visual solutions and knowledge visualization to science instruction and show their mutual influences
- Need to train teachers and instructors to carry on these actions
- Need to motivate and instruct students to come up with their own solutions rather than borrowing and rephrasing what others have done before
- Need to motivate students to get into a habit of sketching on their watch, phone, or tablet on the go
- Need to motivate instructors, teachers, and students to be more playful in their problem-solving tasks
- Apply Howard Gardner strategies and assign people accordingly to fulfill the task
- Enforce collaboration in educational initiatives because people are tired of false competitiveness (of those who want to be seen by others as better than them and the most prominent). Sharing, mutual teaching, inclusivity, and group playfulness should be encouraged.
- Use gender-neutral pronouns and promote ableism-free attitudes and actions as a tool for better collaborative efforts.
- Introduce, in a playful way, coding, chess playing, and languages based on different principles than a child's native tongue, for example, Latin, Japanese, Korean, Arabic, and more. Then, work around it to define theoretical concepts in the curriculum.

References

Ahuja, N., Hwaun, E., Judit R. Pungor, Rafiq, R.,Nemes, S., Sakmar, T., Miranda A. Vogt, Grasse, B., Diaz Quiroz, J., Tessa G. Montague, Ryan W. Null, Danielle N. Dallis, Gavri-ouchkina, D., Marletaz, F., Abbo, L., Daniel S. Rokhsar, Cristopher M. Niell, Soltesz, I., Caroline B. Albertin & Joshua J. C. Rosenthal. (2023). Creation of an albino squid line by CRISPR-Cas9 and its application for *in vivo* functional imaging of neural activity. *Science Direct, Current Biology*. https://doi.org/10.1016/j.cub.2023.05.066

Anderson, A., Krathwohl, D., & Bloom, B. (2000). *A taxonomy for learning, teaching, and assessing: A revision of Bloom's taxonomy of educational objectives*. Retrieved from www. semanticscholar.org/paper/A-Taxonomy-for-Learning%2C-Teaching%2C-and-Assessing% 3A-A-Anderson-Krathwohl/23eb5e20e7985fca5625548d2ee6d781a2861d41

Annie E. Casey Foundation. (2020a). *What the statistics say about generation Z?* Retrieved from www.aecf.org/blog/generation-z-statistics

Annie E. Casey Foundation. (2020b). *What is generation alpha?* Retrieved from www. aecf.org/blog/what-is-generation-alpha

Annie E. Casey Foundation. (2021). *What are the core characteristics of generation Z?* Retrieved from www.aecf.org/blog/what-are-the-core-characteristics-of-generation-z

AP News. (2019, April 24). UN: No screen time for babies; only 1 hour for kids under 5. *Associated Press*. Retrieved from https://apnews.com/article/world-health-organization-health-technology-lifestyle-ap-top-news-407f5d418ab749fd9faa405251071715

App Master: No-code Games: Examples of Projects Created with No Code. (2021). Retrieved from https://appmaster.io/blog/no-code-games-examples-of-projects-created-with-no-code

Baldock, S. J., Punarja, K., Harper, G. R., Griffin, R., Genedy, H. H., James Fong, M., Zhao, Z., Zhang, Z., Shen, Y., Lin, H., Au, C., Jack R. Martin, Mark D. Ashton, Mathew J. Haskew, Stewart, B., Efremova, O., Reza N. Esfahani, Hedley C. A. Emsley, John B. Appleby, Cheneler, D., Damian M. Cummings, Benedetto, A., & John G. Hardy (2023). Creating 3D objects with integrated electronics via multiphoton fabrication *In Vitro* and *In Vivo*. *Advanced Materials Technologies*, *8*(11). Retrieved from https://onlinelibrary. wiley.com/doi/10.1002/admt.202201274

Barroso, A. (2020). Gen Z eligible voters reflect the growing racial and ethnic diversity of U.S. electorate. *Pew Research Center*. Retrieved from www.pewresearch.org/fact-tank/2020/09/23/gen-z-eligible-voters-reflect-the-growing-racial-and-ethnic-diversity-of-u-s-electorate/

BBC The Collection. (2020). *'Cottagecore' and the rise of the modern rural fantasy*. Retrieved from www.bbc.com/culture/article/20201208-cottagecore-and-the-rise-of-the-modern-rural-fantasy.

Beresford Research. (2022). *Generations defined by name, birth year, and ages in 2022*. Retrieved from www.beresfordresearch.com/age-range-by-generation/

Bialik, K., & Fry, R. (2019). Millennial life: How young adulthood today compares with prior generations. *Pew Research Center*. Retrieved from www.pewresearch.org/social-trends/2019/02/14/millennial-life-how-young-adulthood-today-compares-with-prior-generations-2/

bill.com. (2022). Retrieved from www.bill.com/about-us/leadership/rene-lacerte

Biophotonics. (2022). A colophon. *Biophotonics, 29*(4), 3.

Boden, M. A. (2009). Computer models of creativity. *AI Magazine, 30*(3), 23–34.

Boden, M. A. (2012). *Creativity and art: Three roads to surprise*. Oxford University Press.

Borges, J. L. (2006). *The book of imaginary beings*. Peter Sis (Illustrator), Andrew Hurley (Translator). Penguin Classics Deluxe Edition.

Borter, G. (2021). What 'critical race theory' means and why it's igniting debate. *Reuters*. Retrieved from www.reuters.com/legal/government/what-critical-race-theory-means-why-its-igniting-debate-2021-09-21/

Bretous, M. (2022, October 14). What is "quiet quitting"? And why it's trending on social media. *HubSpot*. Retrieved from https://blog.hubspot.com/marketing/quiet-quitting#

Britannica. (2023). *Charles Babbage: British inventor and mathematician*. Retrieved from www.britannica.com/biography/Charles-Babbage

Broadbend, E., Gougoulis, J., Lui, N., Pota, V., & Simons, J. (2017). *Generation Z: Global citizenship survey*. Retrieved from www.varkeyfoundation.org/media/4487/global-young-people-report-single-pages-new.pdf

Brokaw, T. (2001). *The greatest generation*. Random House Trade.

Cain, S. (2013). *Quiet: The power of introverts in a world that can't stop talking*. Crown.

Canva. (2023). Free, printable, customizable storyboard templates. *Canva*. Retrieved from www.canva.com/storyboards/templates/

Cheng, M. (2019, June 19). 8 characteristics of Millennials that support sustainable development goals (SDGs). *Forbes*. Retrieved from www.forbes.com/sites/margueritacheng/2019/06/19/8-characteristics-of-millennials-that-support-sustainable-development-goals-sdgs/?sh=31ce182929b7

Cherry, K., & Susman, D. (2023). Types of nonverbal communication. *Very Well Mind Psychology*. Retrieved from www.verywellmind.com/types-of-nonverbal-communication-2795397

Christopher, N. (2008). *Bestiary*. Dial Press.

COBE. (2023). *NASA science, share the science*. Retrieved from https://science.nasa.gov/missions/cobe

Common Sense: Media. (2023). *Media use by tweens and teens: Infographic*. Retrieved from www.commonsensemedia.org/the-common-sense-census-media-use-by-tweens-and-teens-infographic

Connolly, T., Boyle, E. A., MacArthur, E., Hainey, T., & Boyle, J. M. (2012). A systematic literature review of empirical evidence on computer games and serious games. *Computers & Education, 59*, 661–686.

Cox, Z. N. (2016). Dare to do mighty things: Exploring beyond the earth. In *ACM/SIGGRAPH 2016 Keynote speaker presentation*. Retrieved from http://s2016.siggraph.org/keynote-session

Cross River Therapy. (2022). *Key average attention span statistics*. Retrieved from www.crossrivertherapy.com/average-human-attention-span

Csikszentmihalyi, M. (1996). *The creative personality.* Retrieved September 9, 2022, from www.psychologytoday.com/intl/articles/199607/the-creative-personality

Csikszentmihalyi, M. (1997). *Creativity: Flow and the psychology of discovery and invention.* Harper Perennial.

Csikszentmihalyi, M. (1998). *Finding flow: The psychology of engagement with everyday life* (Masterminds Series). Basic Books.

Daniluk, M., Rocktäschel, T., Welbl, J., & Riedel, S. (2017). *Frustrating short attention spans in neural language modeling* (Conference). ICLR.

Dazed. (2023). AI has now unlocked the ability to read people's minds. *Life & Culturenews.* Retrieved from www.dazeddigital.com/life-culture/article/59778/1/ai-unlocks-the-ability-to-read-people-minds-university-of-texas-research

Deary, I. J., Cox, S. R., & Hill, W. D. (2022). Genetic variation, brain, and intelligence differences. *Molecular Psychiatry, 27*(1), 335–353. Retrieved from https://pubmed.ncbi.nlm.nih.gov/33531661/

Delgado, R., & Stefancic, J. (2023). *Critical race theory* (4th ed.). NYU Press.

Denny, J. (2016, March 3) Growth engineering. *10 Serious Games That Changed The World.* Retrieved from: www.growthengineering.co.uk/10-serious-games-that-changed-the-world/

Denny, J. (2019). The top 10 serious games of all time. *Growth Engineering Ltd.* Retrieved from https://uk.linkedin.com/in/juliettedenny?trk=article-ssr-frontend-pulse_main-author-card

DeVry University. (2023). Retrieved from www.devry.edu/online-programs/area-of-study/media-arts.html

Drevitch, G. (2022). Neuroarts: An emerging field with a plan to transform health. *Psychology Today.* Retrieved from www.psychologytoday.com/us/blog/the-art-effect/202202/neuroarts-emerging-field-plan-transform-health

Drinko, C. (2021). *7 Characteristics of an Aural learner and how they learn best.* Retrieved from www.lifehack.org/883515/aural-learner

Drmanac, R., Andrew B. Sparks, Matthew J. Callow, Aaron L. Halpern, Norman L. Burns, Bahram G. Kermani, Carnevali, P., Nazarenko, I., Geoffrey B. Nilsen, Yeung, G., Dahl, F., Fernandez, A., Staker, B., Krishna P. Pant, Baccash, J., Adam P. Borcherding, Brownley, A., Cedeno, R., Chen, L., Chernikoff, D., Cheung, A., Chirita, R., Curson, B., Jessica C. Ebert, Coleen R. Hacker, Hartlage, R., Hauser, B., Huang, S., Jiang, Y., Karpinchyk, V., Koenig, M., Kong, C., Landers, T., Le, C., Liu, J., Celeste E. McBride, Morenzoni, M., Robert E. Morey, Mutch, K., Perazich, H., Perry, K., Brock A. Peters, Peterson, J., Charit L. Pethiyagoda, Pothuraju, K., Richter, C., Abraham M. Rosenbaum, Roy, S., Shafto, J., Sharanhovich, U., Karen W. Shannon, Conrad G. Sheppy, Sun, M., Joseph V. Thakuria, Tran, A., Vu, D., Alexander Wait Zaranek, Wu, X., Drmanac, S., Arnold R. Oliphant, William C. Banyai, Martin, B., Dennis G. Ballinger, George M. Church & Clifford A. Reid (2009). Human genome sequencing using unchained base. Reads on self-assembling DNA reads on self-assembling DNA nanoarrays. *Science, 327*, 78. https://doi.org/10.1126/science.1181498

The Economist. (2019, February). Generation Z is stressed, depressed and exam-obsessed. *The Economist.* Retrieved from www.economist.com/graphic-detail/2019/02/27/generation-z-is-stressed-depressed-and-exam-obsessed (ISSN 0013–0613).

EPCOT. (2023). Retrieved from https://disneyworld.disney.go.com/disneyworld/epcot

Fitzsimmons, S., Vora, D., Martin, L., Raheem, S., Pekerti, A. A., & Lakshman, C. (2019). Managing yourself: What makes you "multicultural". *Harvard Business Review.* Retrieved from https://hbr.org/2019/12/what-makes-you-multicultural

Freebody, M. (2022, August). More than skin deep: Photonics protects our cultural heritage. *Photonics Spectra*, pp. 54–62. Retrieved from www.photonics.com www.photonics.com/Articles/More_than_Skin_Deep_Photonics_Protects_Our/a68128

Funk, C. (2021). Key findings: How Americans' attitudes about climate change differ by generation, party and other factors. *Pew Research Center*. Retrieved from www.pewresearch.org/fact-tank/2021/05/26/key-findings-how-americans-attitudes-about-climate-change-differ-by-generation-party-and-other-factors/

Gackowski, M. (2015). 3D Rapid Storyboarding – Prototype app overview. *YouTube*. Retrieved from www.youtube.com/watch?v=qQwatDXGttM

GAIA. (2023). Games and intelligent animation. *George Mason University*. Retrieved from https://cs.gmu.edu/~gaia/SeriousGames/index.html

GALA. (2023). *Games and learning alliance conference*. Retrieved from https://conf.seriousgamessociety.org/game-competition/

Gardner, H. (1993/2011). *Art, mind, and brain: A cognitive approach to creativity*. Basic Books, A Division of Harper Collins Publishers.

Gardner, H. (1993/2006). *Multiple intelligences: New horizons in theory and practice*. Basic Books.

Gardner, H. (2000). *Intelligence reframed: Multiple intelligences for the 21st century*. Basic Books.

Gartner. (2023). 9 future of work trends for 2023. *Gartner, Inc.* Retrieved from www.gartner.com/en/articles/9-future-of-work-trends-for-2023

GeeksforGeeks. (2022). *Python Bokeh tutorial – Interactive data visualization with Bokeh*. Retrieved from www.geeksforgeeks.org/python-bokeh-tutorial-interactive-data-visualization-with-bokeh/

Ghasemzadeh, N., & Safari, A. M. (2011). A brief journey into the history of the Arterial Pulse. *SAGE-Hindawi Access to Research Cardiology Research and Practice, 2011*, Article ID 164832, 14 pages. Retrieved from www.hindawi.com/journals/crp/2011/164832/

Gloat, D. (2023). *Study.com. Information visualization: Examples & types*. Retrieved from https://study.com/academy/lesson/information-visualization-examples-types.html

Goh, E. (2019). Gen Zs and Millennials, your learning attention span isn't just 8 seconds. *Smartup Learning Platform*. Retrieved from www.smartup.io/blog/learning/gen-zs-and-millennials-your-learning-attention-span-isnt-just-8-seconds.

Gonzalez, V., Fazlic, I, Cotte, M., Vanmeert, F., Gestels, A., De Meyer, S., Broers, F., Hermans, J., van Loon, A., Janssens, K., Noble, P., & Keune, K. (2023). Lead (II) Formate in Rembrandt's *Night Watch*: Detection and distribution from the macro- to the microscale. Angewandte Chemie International Edition. *Journal of the German Chemical Society*. Retrieved from https://onlinelibrary.wiley.com/doi/10.1002/anie.202216478

Graf, D. (2009). *Point it: Traveller's language kit* (Graf Editions; 16th edition, 2009, 17th edition: ISBN 9783980880275, ASIN B00HTKDAI8). Retrieved from www.amazon.com/Point-Travellers-Language-Original-Dictionary/dp/3980880273/ref=pd_ci_mcx_mh_mcx_views_0?pd_rd_w=Bv5fY&content-id=amzn1.sym.0250fb24-4363-44d0-b635-ac15f859c3b5&pf_rd_p=0250fb24–4363–44d0-b635-ac15f859c3b5&pf_rd_r=CS0E3AX71BJJNQ9VN47Y&pd_rd_wg=rTGKX&pd_rd_r=7f6084d2-b5c6–4ce2–9bf3–43d6bba2d933&pd_rd_i=3980880273

Gregersen, E. (2023). Ada Lovelace: The first computer programmer. *Britannica*. Retrieved from www.britannica.com/story/ada-lovelace-the-first-computer-programmer

Harasim, L. (2017). *Learning theory and online technologies* (2nd ed.). Routledge.

Harvard Business School Online. (2019). *A beginner's guide to data & analytics*. Retrieved from https://online.hbs.edu/Documents/a-beginners-guide-to-data-

and-analytics.pdf?hsCtaTracking=2bb079d4–1f8a-4052–9548–2430ccb52d48%
7C4d888017–3b60–48fb-abd2–754f4106abb4

Henderson, C. (2012). *The book of barely imagined beings.* Granta.

Higher Education Press. (2023, June 27). Integrating visualization with artificial intelligence for efficient data analysis. *Tech Explore.* Retrieved from https://techxplore.com/news/2023-06-visualization-artificial-intelligence-efficient-analysis.html

Holoworld, Artificial Intelligence. (2023). *Embodied artificial intelligence.* Retrieved from www.blogs.holoworld.one/embodied-artificial-intelligence/

Holoworld, The Basics of STREAM Education: A Comprehensive Guide. (2023). Retrieved from www.blogs.holoworld.one/what-is-stream-education/

Huizinga, J. (2016). *Homo Ludens: A study of the play-element in culture.* Angelico Press.

Human Brain Project. (2023). Retrieved from www.humanbrainproject.eu/en/about-hbp/news/events/5055/we-are-science-towards-a-realistic-future-to-enhance-inclusion-gender-equality-and-diversity-in-science

Human Genome Project. (2023). *National Human Genome Research Institute.* Retrieved from www.genome.gov/human-genome-project.

Hutton, J. S., Dudley, J., De Witt, T., & Horowitz-Kraus, T. (2022). Associations between digital media use and brain surface structural measures in preschool-aged children. *Science Reports, 12*(1), 9095. Retrieved from https://pubmed.ncbi.nlm.nih.gov/36351968

IBM. (2023). *What is Natural Language Processing (NLP)?* Retrieved from www.ibm.com/topics/natural-language-processing

Indeed Editorial Team. (2022). 10 common characteristics of the millennial generation. *Indeed.* Retrieved from www.indeed.com/career-advice/interviewing/10-millennial-generation-characteristics

Industry News. (2022, August). Collaboration takes laser welding underwater. *Photonics Spectra,* p. 14. Retrieved from www.photonics.com/Articles/Collaboration_Takes_Laser_Welding_Underwater/a68020

Irvine, K. (2023). XR: VR, AR, MR – What's the difference? *Medium.com.* Retrieved from https://medium.com/viget-collection/xr-vr-ar-mr-whats-the-difference-ac7c3fcd590d

Isaacson, W. (2008). *Einstein: His life and universe.* Simon and Schuster.

Isaacson, W. (2021). *The code breaker: Jennifer Doudna, gene editing, and the future of human race.* Simon & Schuster.

IVLA, International Visual Literacy Association. (2023). Retrieved from https://web.archive.org/web/20031204235851/http://ivla.org/org_what_vis_lit.htm

James, A. V., James, D. A., & Kalisperis, L. N. (2004). A unique art form: The friezes of Pirgi. *Leonardo, 37*(3), 234–242.

JCSG. (2022). *Conference proceedings, serious games joint international conference, Weimar, Germany.* Retrieved from https://link.springer.com/book/10.1007/978-3-031-15325-9

JCSG. (2023). *International annual joint conference on serious games.* Retrieved from https://jointconference-on-seriousgames.org/
www.jneurosci.org/content/32/48/17492

Kasanoff, B. (2017). Intuition is the highest form of intelligence. *Forbes.* Retrieved from www.forbes.com/sites/brucekasanoff/2017/02/21/intuition-is-the-highest-form-of-intelligence/?sh=4fbfe8638602

Kaufman, K. (2019). *A season on the wind: Inside the world of spring migration.* Houghton Mifflin Harcourt.

Kent, S. (2023). *With fur out of fashion, PETA sets its sights on wool, leather and down.* Retrieved from www.cnn.com/style/peta-wool-leather-down-bof/index.html

Kneeland, R., Ojeda, J., St-Yves, G., & Naselaris, T. (2023). *Reconstructing seen images from human brain activity via guided stochastic search*. Retrieved from https://arxiv. org/abs/2305.00556

Koch, C. (2015). Intuition may reveal where expertise resides in the brain. *Scientific American*. Retrieved from www.scientificamerican.com/article/intuition-may-reveal-where-expertise-resides-in-the-brain/

Kolbert, E. (2023). Elemental need. *The New Yorker, XCIX*(3), 24–27.

Kosowatz, J. (2020). Three advances in prosthetics. *The American Society of Mechanical Engineers*. Retrieved from www.asme.org/topics-resources/content/three-advances-in-prosthetics

Kyriacou, P., & Abay, T. Y. (2019). Optical technology. In R. Narayan (Ed.), *Encyclopedia of biomedical engineering*. Elsevier Inc.

Lakoff, G. (1990). The invariance hypothesis: Is abstract reason based on image-schemas? *Cognitive Linguistics, 1*(1), 39–74.

Lakoff, G., & Johnson, M. (1980/2003). *Metaphors we live by*. The University of Chicago Press.

Lakoff, G., & Núñez, R. E. (2001). *Where mathematics comes from: How the embodied mind brings mathematics into being*. Basic Books.

LaMotte, S. (2019). MRIs show screen time linked to lower brain development in preschoolers. *CNN Health*. Retrieved from www.cnn.com/2019/11/04/health/screen-time-lower-brain-development-preschoolers-wellness/index.html

Lander, E. S. (2016). The heroes of CRISPR. *Cell, 164*, 18–28. Retrieved from www.cell. com/cell/fulltext/S0092-8674(15)01705-5?_returnURL=https%3A%2F%2Flinking hub.elsevier.com%2Fretrieve%2Fpii%2FS0092867415017055%3Fshowall%3Dtrue

Laning, T. (2020). *Grendel Games*. Retrieved from https://grendelgames.com/serious-games-gamificationand-game-based-learning-whats-the-difference/

LePan, N. (2020). Visualizing the history of pandemics. *World Economic Forum*. Retrieved from www.weforum.org/agenda/2020/03/a-visual-history-of-pandemics/

Lillemyr, O. F. (2020). *Taking play seriously: A challenge of learning* (2nd ed.). Information Age Publishing.

Lima, M. (2017). *The book of circles: Visualizing spheres of knowledge (with over 300 beautiful circular artworks, infographics and illustrations from across history)*. Princeton Architectural Press.

Lima, M., & Shneiderman, B. (2014). *The book of trees: Visualizing branches of knowledge*. Princeton Architectural Press.

Lin, C.-H., Liu, E. Z.-F., Chen, Y.-L., Liou, P.-Y., Chang, M., Wu, C.-H., & Yuan, S.-M. (2013). Game-based remedial instruction in mastery learning for upper-primary school students. *Educational Technology & Society, 16*(2), 271–281.

Lotek. (2023). *Monitor all your birds and bats simultaneously with 0.13g coded radio tags*. Retrieved from www.lotek.com/products/nanotags/

Lugmayr, A., Suhonen, J., Hlavacs, H., Montero, C., Suutinen, E., & Sedano, C. (2016). Serious storytelling – A first definition and review. *Multimedia Tools and Applications, 76*(14), 15707–15733. Retrieved from www.academia.edu/31528082

Main, D. (2017). Who are the millennials? *Live Science*. Retrieved from www.livescience. com/38061-millennials-generation-y.html

Malraux, A. (1958/1974). *La Métamorphose des Dieux*. Gallimard.

Malraux, A. (1996). *Le musée imaginaire*. Gallimard.

Marchant, J. (2020). *The human cosmos: Civilization and the stars*. Dutton.

Marshack, A. (1972). *The roots of civilisation*. Littlehampton Book Services Ltd.

McCrindle, M. (2022). *The power of visuals to tell a story*. McCrindle Research Pty Ltd. Retrieved from https://mccrindle.com.au/service/design/the-power-of-visuals-to-tell-a-story/

McNerney, S. (2011). A brief guide to embodied cognition: Why you are not your brain. *Scientific American*. Retrieved from https://blogs.scientificamerican.com/guest-blog/a-brief-guide-to-embodied-cognition-why-you-are-not-your-brain/

Microsoft Bing. (2023). *Starwars fighter*. Retrieved from www.bing.com/images/search?view=detailV2&ccid=lsDZXLS5&id=E49C32D5B900599BB58D366604B1A6CA DD961DBE&thid=OIP.lsDZXLS5qxPdbqcwFntGRQHaFj&mediaurl=https%3a%2f%2fwallpapercave.com%2fwp%2flbHWVho.jpg&cdnurl=https%3a%2f%2fth.bing.com%2fth%2fid%2fR.96c0d95cb4b9ab13dd6ea730167b4645%3frik%3dvh2W3cqm sQRmNg%26pid%3dImgRaw%26r%3d0&exph=768&expw=1024&q=Ty+Plane+St ar+Wars&simid=607986315312630534&FORM=IRPRST&ck=5C0514C5C62104A E4D2F2BE39AD5D05D&selectedIndex=1&ajaxhist=0&ajaxserp=0as I

Miller, K. (2019).17 Data visualization techniques all professionals should know. *Harvard Business School Online*. Retrieved from https://online.hbs.edu/blog/post/data-visualization-techniques

Mok, A., & Zinkula, J. (2023). *ChatGPT may be coming for our jobs. Here are the 10 roles that AI is most likely to replace*. Retrieved from www.businessinsider.com/chatgpt-jobs-at-risk-replacement-artificial-intelligence-ai-labor-trends-2023–02

Moro, C., Stromberga, Z., & Birt, J. R. (2020). Technology considerations in health professions and clinical education. In D. Nestel, G. Reedy, L. McKenna, & S. Gough (Eds.), *Clinical education for the health professions: Theory and practice* (p. 25). Springer Nature Singapore Pte Ltd. https://doi.org/10.1111/medu.14251

Morrison-Williams, S. (2022). Millennials — Changing the Face of higher education. Education initiative. *The Pacific Institute*. Retrieved from https://educationinitiative.thepacificinstitute.com/articles/story/millennials-changing-the-face-of-higher-education

Motus: Wildlife Tagging System. (2023). *Tag deployment*. Retrieved from https://motus.org/tag-deployment/

Nahas, K. (2023). AI re-creates what people see by reading their brain scans. *Science News*. Retrieved from www.science.org/content/article/ai-re-creates-what-people-see-reading-their-brain-scans

Nalven, J. (2023). The future of AI? *Minding the Campus*. Retrieved from www.mindingthecampus.org/2023/07/19/the-future-of-ai/

NASA. (2023). Retrieved from www.nasa.gov/

NASA Image Galleries. (2023). Retrieved from www.nasa.gov/multimedia/imagegallery/index.html

National Research Council of the National Academies. (2008). *Inspired by biology: From molecules to materials to machines*. The National Academies Press.

NYU Game Center. (2023). Retrieved from https://gamecenter.nyu.edu/

O'Hare, W., & Mayol-Garcia, Y. H. (2023). The changing child population of the United States: first data from the 2020 census. *Annie E. Casey Foundation*. Retrieved from https://assets.aecf.org/m/resourcedoc/aecf-changingchildpop-2023.pdf#page=3

Patel, D. (2017). 10 tips for marketing to Gen Z consumers. *Forbes*. Retrieved from www.forbes.com/sites/deeppatel/2017/05/01/10-tips-for-marketing-to-gen-z-consumers/?sh=a3978df3c503

Pepperberg, I. M. (1999). *The Alex studies*. Harvard University Press.

Pepperberg, I. M. (2008). *Alex & me*. Harper-Collins Publishers.

Pepperberg, I. M. (2017). Symbolic communication in nonhumans. In J. Call, I. M. Pepperberg, C. T. Snowdon, & T. R. Zentall (Eds.), *APA handbook of comparative psychology*. APA Press.

Prisco, G. (2018). Gitanjali Rao wants to make polluted water safer with lead detection system. *CNN Health*. Retrieved from https://edition.cnn.com/2017/11/28/health/gitanjali-rao-young-scientist-winner/index.html

Putz, M. V. (2022). *Special issue symmetries in quantum nano-chemistry (From structure to properties, observability and functions)*. Retrieved from www.mdpi.com/journal/symmetry/special_issues/Symmetries_Quantum_Nano-Chemistry_Structure_Properties_Observability_Functions

Quellette, J. (2023). Scientists identify rare lead compounds in Rembrandt's. *The Night Watch. Art Technica*. Retrieved from https://arstechnica.com/science/2023/01/scientists-identify-rare-lead-compounds-in-rembrandts-the-night-watch/

Quiles, A., Valladas, H., Bocherens, H., Delque-Kolic, E., Kaltnecker, E., van der Plicht, J., Delannoy, J. J., Feruglio, V., Fritz, C., Monney, J., Philippe, M., Tosello, G., Clottes, J., & Geneste, J.-M. (2016). A high-precision chronological model for the decorated upper palaeolithic cave of Chauvet-Pont D'arc, Ardeche, France. *Proceedings of the National Academy of Sciences of the United States of America, 113*(17), 4670–4675.

Rao, G. (2021). *A young innovator's guide to STEM: 5 steps to problem solving for students, educators, and parents*. Post Hill Press.

Reche, I., & Perfectti, F. (2020). Promoting individual and collective creativity in science students. *Trends in Ecology & Evolution, 35*(9), 745–748.

Researchers Adapt Laser Welding for 3D-Printed Parts. (2021, August). *Photonics spectra* 2021. Retrieved from www.photonics.com/Articles/Researchers_Adapt_Laser_Welding_for_3D-Printed/a67040

Robbin, T. (2006). *Shadows of reality: The fourth dimension in relativity, cubism, and modern thought* (1st ed.). Yale University Press.

Ross, S. T., Schwartz, S., Fellers, T. J., & Davidson, M. W. (2022). Total internal reflection fluorescence (TIRF) microscopy. *Nikon*. Retrieved from www.microscopyu.com/techniques/fluorescence/total-internal-reflection-fluorescence-tirf-microscopy

Rothman, J. (2023). How should we think about our different styles of thinking? *The New Yorker*. Retrieved from www.newyorker.com/magazine/2023/01/16/how-should-we-think-about-our-different-styles-of-thinking

Ryssdal, K. (2014). Goldfish have longer attention spans than Americans, and the publishing industry knows it. *Marketplace*. Retrieved from https://internet.psych.wisc.edu/wp-content/uploads/532-Master/532-UnitPages/Unit-09/Attention_Goldfish_Abbreviated.pdf.

Safina, C. (2015). *Beyond words what animals think and feel*. Henry Holt and Company.

Scheffer, M. (2014). The forgotten half of scientific thinking. *Proceedings of the National Academy of Sciences of the United States of America, 111*, 6119.

Schlett, J. (2022, August). STEM programs struggle to satisfy the 'endless demand' for photonics talent. *Photonics Spectra*, pp. 43–49. Retrieved from www.photonics.com www.photonics.com/Articles/STEM_Programs_Struggle_to_Satisfy_the_Endless/a68145

Science History Institute. (2022). *Francis Crick, Rosalind Franklin, James Watson, and Maurice Wilkins*. Retrieved from www.sciencehistory.org/historical-profile/james-watson-francis-crick-maurice-wilkins-and-rosalind-franklin

Secher, J. (2022, September). Loveland hosts famous child scientist at innovation celebration. *NOCO Style, Northern Colorado's lifestyle Magazine*. Retrieved from https://nocostyle.com/2022/09/15/loveland-hosts-famous-child-scientist-at-innovation-celebration/

Shankar, R. (2022, September 14). Why quiet quitting is actually a good thing for your employees and your business. *Do Change Right*. Retrieved from https://dochangeright.com/why-quiet-quitting-is-actually-a-good-thing-for-your-employees-and-your-business/

Shankland, S. (2016). *Minecraft: The video game that builds kids' brain cells*. Retrieved from www.cnet.com/tech/gaming/features/minecraft-the-video-game-that-builds-kids-brain-cells/

Sliwa, J. (2018). Teens today spend more time on digital media, less time reading. *American Psychological Association.* Retrieved from www.apa.org/news/press/releases/2018/08/teenagers-read-book.

Southern New Hampshire University. (2023). Retrieved from https://degrees.snhu.edu/programs/bs-in-game-programming-and-development

Spitznagel, E. (2020, January 25). Generation Z is bigger than millennials — And they're out to change the world. *New York Post.* Retrieved from https://nypost.com/2020/01/25/generation-z-is-bigger-than-millennials-and-theyre-out-to-change-the-world/

Statistic Brain. (2018). Retrieved May 20, from www.statisticbrain.com/attention-span-statistics/

STEAM. (2023). *Rhode Island school of design.* Retrieved from www.risd.edu/steam.

STEM to STEAM: Integrated Studies STEM to STEAM Resources Toolkit. (2023). *Edutopia, George Lucas Educational Foundation.* Retrieved from www.edutopia.org/stem-to-steam-resources

Sternberg, R. J. (1986/2023). A triangular theory of love. *Psychological Review, 93*(2), 119–135. Retrieved from https://psycnet.apa.org/record/1986-21992-001

Studiobinder. (2023). *Download a FREE storyboard template for word or make a storyboard online.* Retrieved from www.studiobinder.com/blog/downloads/storyboard-template-word/

Sullivan, W. (2022). By reading brainwaves, an A.I. aims to predict what words people listened to. *Smithsonian Magazine.* Retrieved from www.smithsonianmag.com/smart-news/by-reading-brain-waves-an-ai-could-predict-what-words-people-listened-to-180980738/

Sweatman, M. B. (2018). *Cave paintings reveal use of complex astronomy.* The University of Edinburg. Retrieved from www.ed.ac.uk/news/2018/cave-paintings-reveal-use-of-complex-astronomy

Sweatman, M. B. (2020). Zodiacal dating prehistoric artworks. *Athens Journal of History, 6*(3), 199–222. Retrieved from https://doi.org/10.30958/ajhis.6-3-2

Sweatman, M. B., & Coombs, A. (2020). Decoding European palaeolithic art: Extremely ancient knowledge of precession of the Equinoxes. *Athens Journal of History, 5*(1), 1–30.

Takagi, Y., & Nishimoto, S. (2023). High-resolution image reconstruction with latent diffusion models from human brain activity. *bioRxiv Preprint.* Retrieved from www.science.org/

Tang, J., LeBel, A., Jain, S., et al. (2023). Semantic reconstruction of continuous language from non-invasive brain recordings. *Nature Neuroscience, 26,* 858–866. https://doi.org/10.1038/s41593-023-01304-9

Taras, V., Baack, D., Caprar, D., Jimenéz, A., & Froese, F. (2021, June 9). Research: How cultural differences can impact global teams. *Harvard Business Review.* Retrieved from https://hbr.org/2021/06/research-how-cultural-differences-can-impact-global-teams?ab=at_art_art_1x4_s03

Thomas, M. (2011). *Deconstructing digital natives: Young people, technology, and the new literacies.* Taylor & Francis.

Tongwaranan Tanyatorn. (2019, August 12). The power of 8 seconds. *Bangkok Post.* Retrieved from www.bangkokpost.com/business/1728799/the-power-of-8-seconds.

Trafton, A. (2023). Using AI, scientists find a drug that could combat drug-resistant infections. *MIT News.* Retrieved from https://news.mit.edu/2023/using-ai-scientists-combat-drug-resistant-infections-0525

Turner, A. (2023). Everyone's talking about "quiet quitting." Here's what it means — And how the term got its start. *Business Insider.* Retrieved from www.businessinsider.com/quiet-quitting-coasting-nickelodeon-dan-schneider-shopify-slack-2022-9

UNICEF, Office of Innovation. (2020). *Kindly*. Retrieved from www.unicef.org/innovation/kindly

Valladas, H., H. Cachier, P. Maurice, F. Bernaldo de Quirost, J. Clottes, V. Cabrera Valdés, P. Uzquiano & M. Arnold (1992). Direct radiocarbon-dates for prehistoric paintings at the Altamira, El-Castillo and Niaux Caves. *Nature*, *357*, 6373.

Valladas, H., Kaltnecker, E., Quiles, A., Tisnérat-Laborde, N., Genty, D., Arnold, M., E. Delqué-Kolic, C. Moreau, Baffier, D., J. J. Cleyet Merle, Clottes J., Girard, M., Monney, J., R. Montes, González Sainz, C., Sanchidrián Torti, J. L., & Simonnet, R. (2013). Dating French and Spanish prehistoric decorated caves in their archaeological context. *Radiocarbon*, *55*, 2–3.

Valladas, H., Quiles, A., Delqué-Kolic, & Kaltnecker, E. (2017). Radiocarbon dating of the Decorated Cosquer Cave (France). *Radiocarbon*, *59*(2) 1–13. https://doi.org/10.1017/RDC.2016.87

Van Gulch, R. (2014). Consciousness. *Stanford Encyclopedia of Philosophy*. Retrieved from https://plato.stanford.edu/entries/consciousness/

VisionShiftStudios. (2023). Retrieved from www.visionshiftstudios.com/

W3C Web Accessibility Initiative (WAI). (2023). Retrieved from www.w3.org/WAI/

Wai, J., Lakin, J. M., & Kell, H. J. (2022). Specific cognitive aptitudes and gifted samples. *Intelligence*, *92*, 101650. Retrieved from www.sciencedirect.com/science/article/abs/pii/S0160289622000319

Wan, X., Taken, D., Asamizuya, T., Suzuki, C., Ueno, K., Cheng, K., Ito, T., & Tanaka, K. (2012). Developing intuition: Neural correlates of cognitive-skill learning in caudate nucleus. *Journal of Neuroscience*, *32*(48), 17492–17501. https:.org/10.1523/JNEUROSCI.2312-12.2012

Wikipedia. (2023). List of simulation video games. *Wikipedia*. Retrieved from https://en.wikipedia.org/wiki/List_of_simulation_video_games

Wilson, L. O. (2020). *Three domains of learning – Cognitive, affective, psychomotor*. Retrieved from https://thesecondprinciple.com/instructional-design/threedomainsoflearning/

Worrall, S. (2015, July 15). Yes, animals think and feel. Here's how we know. *National Geographic*. Retrieved from https://news.nationalgeographic.com/2015/07/150714-animal-dog-thinking-feelings-brain-science/

Yakman, G. (2008). *STEAM education: An overview of creating a model of integrative education*. Retrieved from www.researchgate.net/publication 327351326_STEAM_Education_an_overview_of_creating_a_model_of_integrative_education?enrichId=rgreq-3c1f1782cb441cece1183488e9b2b655-XXX&enrichSource=Y292ZXJQYWd-1OzMyNzM1MTMyNjtBUzo2NjU4NDAwNTk5NTMxNTJAMTUzNTc2M-DA0OTkzMA%3D%3D&el=1_x_3&_esc=publicationCoverPdf

Yakman, G., & Lee, H. (2012). Exploring the exemplary STEAM education in the U.S. as a practical educational framework for Korea. *Journal of The Korean Association for Science Education, 32*(6), 1072–1086.

Zauderer, S. (2022). Average human attention span by age (Infographic). *Cross River Therapy*. Retrieved from www.crossrivertherapy.com/average-human-attention-span

Zheng, R., & Gardner, M. (Eds.). (2017). *Handbook of research on serious games for educational applications*. IGI Publishing.

Zoupan, C. T. (1760). De pulsuum differentiis simplicibus. *Stanno Schneideriano, Halae Magdeburgicae*. Retrieved from https://books.google.com/books/about/De_pulsuum_differentiis_simplicibus.html?id=-fW0GwAACAAJ

Glossary: Terms and definitions

Abstract – a concept or thought that does not have physical, concrete existence. Also, an artistic object that has a meaning independent of narrative or pictorial representation. Also, a summary of a larger text.

Aesthetics – a philosophical analysis of principles related to beauty, especially in the arts

Algorithm is a mathematical sequence of instructions telling how to perform the computation to make a program. Algorithms are often used to create repetition by applying an algorithm multiple times; it is a recursive process. Algorithms serve to solve a complex problem by writing a sequence of more simple, unambiguous steps. Such a course of action is used for writing computer programs and in programmed learning.

Anthropology means the study of human characteristics. Physical anthropology studies human biological and physiological features and their evolution. Cultural anthropology studies societies, cultures, and their development.

API is an application programming interface, a software interface that allows communication between computers and/or programs.

Artifact – an object produced by human craft (a tool, a weapon, or an ornament of archaeological or historical interest). Thus, Duchamp's *Fountain* may be both an artwork and an artifact.

Augmented reality (AR) is an overlay of computer-generated content on the real world that can superficially interact with the environment in real time. With AR, the CG content and the real world do not block themselves.

Binary numeral system uses two symbols, typically 0 (zero) and 1 (one).

Biointerface is a place of contact between a biomolecule, living tissue, or organism and a living or another material. Nanostructure interfaces are extensively studied.

ChatGPT is a computer program that uses artificial intelligence (AI) and natural language processing (NLP) to understand customer questions and automate responses to them, simulating human conversation (*ibm.com/ topics/chatbots*).

Cognitive abilities   make it possible to think abstractly (not only on concrete objects), learn from experience, reason, comprehend ideas and concepts, and solve problems. They support gaining knowledge, utilizing information, and creative and critical thinking.

Cobots   are robots that collaborate with people.

DALL·E 2   is a version of GPT-3 that can create realistic images and art from text descriptions.

Data   – factual information specially organized for analysis, reasoning, or making decisions

Data visualization   – information abstracted in a schematic form to provide visual insights into data sets. Data visualization enables us to go from the abstract numbers in a computer program (ones and zeros) to the visual interpretation of data. Text visualization means converting textual information into graphic representation so we can see the information without reading the data, such as tables, histograms, pie or bar charts, or Cartesian coordinates.

Dimension   measures spatial extent, width, height, or length. Dimension means how many numbers (coordinates) are needed to determine a position of a point in space. In the Cartesian three-dimensional space, three coordinates, x, y, and z, tell about the position of a point: x-horizontal axis, y-vertical axis, and z-depth. (Cartesius (whose French name was René Descartes) revolutionized mathematics by providing the first systematic link between geometry described by Euclid and algebra.) A **point** has no length, area, volume, or other dimensional attributes, so some mathematicians claim a point is 0-dimensional. A **line** has one dimension because only one coordinate is needed to describe a point in space, while a **surface** has two dimensions. Data may be linear or multidimensional.

Electromagnetic radiation   results from the electromagnetic field's propagation of waves (or photons), which carry radiant energy. Waves, which have different wavelengths, include radio waves, microwaves, infrared, light (which is visible), ultraviolet light, X-rays, and gamma rays.

Empathy   means understanding and sharing another person's or creature's feelings. One may apply cognitive (by knowing how to take somebody's perspective), emotional (by feeling somebody's emotions), and compassionate (by combining intellect and feelings) empathy.

Extended reality (XR)   comprises real-and-virtual environments generated by computer technology and wearables.

Fiber lasers   use fiber-optic cables and cut stones, metals, glass, and most other materials with a precision of about 1/4 of the diameter of human hair.

Form   – mass that occupies space, giving the weight, density, and thickness of a 3D object or implied in a picture of this object

Fractal geometry   describes self-similar or scale-symmetric objects called fractals. When magnified or reduced in size, fractal things are self-similar: their parts are similar to the whole, and the likeness continues when the parts are magnified more and more. They are ragged at every scale, less smooth than Euclidean lines, planes, and spheres. Fractals are present in the design and artwork. Fractal geometry can explain the origination of patterns in living organisms.

GPT (generative pretrained transformer) is a deep learning model originated by Google in 2017.

GPT-4 is a multimodal model that accepts prompts of images and texts as inputs and gives out text outputs, such as natural language or code.

Graphic is an image represented by a graph or relating to graphics. Computers often generate graphic displays.

Haptic relates to the sense of touch; the senses of touch and proprioception enable the perception and manipulation of objects.

Higher-order thinking goes beyond basic observation of facts and data memorization, enabling you to build data-based knowledge. Higher-order thinking makes critical thinking possible in an evaluative, creative, and innovative way.

HTML – Hypertext Markup Language. This language codes formatting, links, and other things on Web pages. **XML** – eXtensible Markup Language. A markup language like HTML; it lets individuals define and use their tags.

Hyperlink – a link from a hypertext file or document to another file activated by clicking on a highlighted word or image on the screen

Icon, iconic object, or image – An icon represents a thing or refers to something by resembling or imitating it; thus, a picture, a photograph, a mathematical expression, or an old-style telephone may be regarded as an iconic object. Therefore, an iconic object or image has qualities in common with things it represents by looking, sounding, feeling, tasting, or smelling alike.

Iconography means visual elements such as images and symbols used in a specific culture, society, or artwork. It also may indicate study of these elements.

Immersive experience happens when a 3D or 4D (including time) image (maybe plus sound or other stimuli) surrounds you, creating extended reality. Platforms that support immersive experiences include augmented reality (AR), virtual reality (VR), mixed reality (MR), 360° video, wearable devices, the Internet of Things (IoT), and more.

Infographics – tools and techniques involved in the graphical representation of data, mostly in journalism, art, and storytelling

Information – knowledge derived from study, experience, or instruction based on facts or data

Information visualization – representation plus interaction – means using the computer-supported, visual-spatial representation of abstract data to amplify cognition and derive new insights. Data presented as information visualization are often interactive, numerical, verbal, and graphical (such as the text of geographic representations).

Ions: cations and anions   – An atom or molecule with an electric charge is called an ion. An electric charge results from the presence of single, double, triple, or even higher negative electrons, unequal to the number of positive protons in the nucleus of an atom. Removing or adding one or more electrons changes a neutral atom into an ion. A cation is an ion or group of ions with a positive charge. An anion is a negatively charged ion. In an electrolyte solution, cations move toward the cathode (negative electrode) after an electric current is applied, and anions migrate to an anode.

Interface   denotes a connection or interaction between two units (which may be subjects, organizations, or physical or electronic systems), for example, social interface or interface between phases of matter such as solid, liquid, and gas, and also interaction of hardware and software. The user–computer interface allows communication with the operating system. Interface with the internet may go through the internet protocol.

Internet of Things   (IoT) means a set of objects connected over a wireless network and acting without human actions. It serves for personal use in smart homes (with everyday physical devices embedded with sensors and software), for care of elders with disabilities, as an Internet of Medical Things (IoMT), for business, industrial, vehicular communication, and many kinds of infrastructure systems.

Kinesthesia   means knowing about the position and movement of the parts of the body. Sensory organs (proprioceptors) in the muscles and joints signal changes in position.

Knowledge visualization   uses visual representations to transfer insights and create new *knowledge* in communicating different visual formats.

Latitude and longitude   are parts of a grid system, giving us horizontal and vertical coordinates in the geographic coordinate system. Latitude lines run east–west, and longitude lines run north–south. Every location on Earth has a global address described by two numbers called latitude and longitude coordinates; they say where the coordinates intersect.

LED,   the light-emitting diode, is a semiconductor diode that glows when a voltage is applied.

Mandala   is a geometric circular figure depicting the universe according to Hindu and Buddhist traditional symbols.

Manifesto   is a public declaration of opinions, statements, and programs, mostly of political or social aims. An art manifesto expresses aesthetic and philosophical thoughts about the art defined by an artist or a group of artists. An individual art manifesto facilitates the translation of artistic, visual messages into literary form and conveys it to viewers, critics, and jurors.

Meander   is a geometric pattern made by a line bending along and repeatedly crossing a straight line but not crossing itself. In the arts, it is a decorative pattern constructed from a continuous line shaped into a repeated, ornamental motif.

Metaphor – indicates one thing as representing another, thus making an implied mental comparison. For example, we use metaphors every day when we talk about desktop items, such as folders, documents, and in- and outboxes, using familiar objects for organizing elements related to the computer. Also, a figure of speech where we use one word or phrase to designate another one, thus suggesting comparison, as in 'a sea of trouble.'

Mime – a performance art involving body motions without the use of speech

Mixed reality (MR) – In MR, the synthetic content is overlaid and anchored to and interacts with objects in the real world in real time.

Mythology is a collection of stories, myths, legends, and beliefs typical of a given culture or religion; for example, Christian, Jewish, Egyptian, Hindu, or Muslim mythologies.

NASA is the National Aeronautics and Space Administration, a civil space program and space exploration. This agency works with contractors, academia, and international and commercial partners to explore, discover, and expand knowledge for the benefit of humanity.

Negative space is the space around the objects depicted in a picture. A **figure–ground reversal** is an optical illusion that happens when we perceive an image formed by the negative space.

Network is a set of connected (wired or wireless) devices, such as computers, servers, mainframes, peripherals, firewalls, smartphones, and more, which allows people to share data. Devices may be linked through cables, telephone lines, radio waves, satellites, or infrared light beams.

Networking means exchanging ideas and information among people for social or business contact.

Neurotransmitters are chemical substances released by nerve cells that transfer signals from neurons to other cells.

OpenAI is a research unit that derives its solutions from deep learning and natural language processing.

Pantomime – a musical, theatrical production using gestures and movements but not words

Paradigm is a pattern of thought: a set of concepts, or a way of thinking characteristic of a specific field in science (research methods, postulates, and standards) or philosophy. A paradigm shift means transitioning to a new approach to the field under study and is how developments come about. Paradigm also means a model, a typical example.

Pattern means the regular order existing in nature and a manmade design. We see patterns everywhere in nature, mathematics, art, architecture, and design. In nature, patterns can be seen as symmetries (e.g., snowflakes) and/or structures having fractal dimensions such as spirals, meanders, or surface waves. In computer science, design patterns serve in creating computer programs. In the arts, the pattern is an artistic or decorative design made of recurring lines or any repeated elements.

Perspective is the appearance of depth, picturing the world on a two-dimensional, flat surface of a piece of paper or a computer screen. Perspective represents 3D objects and depth relationships on a 2D surface.

Photonics is the science of light (photons) generation, detection, and manipulation. It can go through emission, transmission, modulation, sensing, signal processing, switching, and amplification.

Scale is a range of values that forms a standard system for measuring the size of something. Units for measuring in science are known as the International System of Units, which is based on the meter. Very small physical and biological structures are measured fractions of a meter: micrometer (one-millionth of a meter − 1×10^{-6} m), nanometer (one billionth of a meter − 1×10^{-9} m), and picometer (one trillionth of a meter − 1×10^{-12} m). Thus, the picometer is one-thousandth of a nanometer. Below are examples of objects of various sizes.

The Milky Way galaxy (about 9.5×10^{17} km in diameter. There are 100–400 billion (1–4) $\times 10^{11}$) stars.
The Earth (10^{18} m)
A city (10^{4} m)
A tree (10^{1} m)
A leaf (10^{-1} m)
A cell (10^{-5} m)
Strands of DNA (10^{-7} m)
An atom (10^{-10} m), and
Quarks (10^{-16} m)

Scientific visualization presents real, abstract, or model-based objects in a digital way directly from the data. It may deliver art–science cooperative learning projects and make knowledge understandable to a broad audience. Visualization as storytelling comprises narratives, interactive graphics, explanatory and animated graphics, and multimedia.

Semiconductor material has a conductivity between a conductor (most metals, for example, copper) and an insulator (such as glass). The properties of a semiconductor, its moving electrons, and electron holes in a crystal lattice can be explained by quantum physics. Silicon crystals are semiconductive materials; they are used in microelectronics and photovoltaics used for direct conversion of sunlight to electricity in solar panels.

Semiotics studies the meaningful use of signs, symbols, codes, and conventions for communication. The name 'semiotics' is derived from the Greek word 'semeion,' which means 'sign.' 'Meaning' is always the result of social conventions, even when we think something is natural or characteristic and use signs for those meanings. Therefore, culture and art are series of sign systems. Semioticians analyze such sign systems in various cultures; linguists study language as a system of signs, and some even examine film as a system of signs. The semiotic content of visual design is essential for nonverbal communication applied to practice, especially for visualizing knowledge.

Sign tells about a fact, an idea, or information; it is a distinct thing that signifies another thing. Natural signs signify events caused by nature, while conventional signs may signal art, social interactions, fashion, food, interaction with technology, machines, and practically everything else. Signs take conventional shapes or forms to tell about facts, ideas, or information. Icons and symbols help visually compress information. An **icon** represents a thing or refers to something by resembling or imitating it; thus, a picture, a photograph, a mathematical expression, or an old-style telephone may be regarded as an iconic object. Therefore, an **iconic object** has qualities in common with things it represents by looking, sounding, feeling, tasting, or smelling alike. Natural signs signify events caused by nature, while conventional signs may signal art, social interactions, fashion, food, interactions with technology, machines, and practically everything else.

Signage is a visual graphic that displays information, for example, street signs, room identification signs, or any informational or regulatory signs. Signs, symbols, and icons are collectively called signage. Effective design of a complicated product may help people memorize and learn how to use this product (for example, 'Where is the switch?' or 'How do I open this thing?').

Social media include social networking sites, image-sharing sites, video-hosting sites, discussion sites, community blogs, and economic networks.

Social networking means using internet-based social media websites and applications to interact with other users for a social or business purpose and stay connected with friends, family, colleagues, customers, or clients. **Social networking** acts through sites like Facebook, X (formerly called Twitter), LinkedIn, and Instagram.

Space is a three-dimensional expanse in the physical universe where we can measure the position and direction of objects and events. A Cartesian coordinate system (with the x, y, and z axes) serves this purpose in Euclidean geometry.

Spectroscopy records how matter interacts with or emits electromagnetic radiation. A spectroscope serves to study matter by measuring properties of specific frequencies and wavelengths of electromagnetic radiation.

Symbol does not resemble things it represents but refers to something by convention; for example, the word 'red' represents red. We must learn the relationship between symbols and what they represent, such as letters, numbers, words, codes, traffic lights, and national flags. A symbol represents an abstract concept, not just a thing, and is comparable to an abstract term. Highly abstracted drawings that show no realistic graphic representation become symbols. Symbols are omnipresent in our lives, for example:

- An electric diagram that uses abstract symbols for a light bulb, wire, connector, resistor, and switch

- An apple for a teacher or a bitten apple for a Macintosh computer
- A map – a typical abstract graphic device
- A 'slippery when wet' sign

Symmetry is present when similar parts of an object are arranged around a line, point, or plane. A crystal shows symmetry when it has a center of symmetry, rotation axes, or mirror planes (imaginary planes that divide it into halves). There are several types of symmetry: for example, line or mirror symmetry, radial, cylindrical, or spherical symmetry. A figure with a line of bilateral symmetry has two identical halves when folded along its line of symmetry, and these halves are congruent, meaning they are the same size and shape. An object has radial symmetry when it can be rotated around the rotation axis. For example, with a fourfold rotation axis, the crystal repeats itself each 90°. Angles of rotational symmetry possible for crystals are: 60°, 90°, 120°, 180°, and 360°. The halves of the bilaterally symmetrical animals, for example, butterflies, form each other's mirror images when seen along the axis. A circle is symmetrical about its center. Optical isomers are 'mirror images' of each other and are symmetrical around a plane. For example, molecules of some sugar isomers deflect the rays of light in the right or left direction.

Synesthesia means the merging of two or more senses. It happens when a sensory stimulus is received by one sense but also triggers perception in another sense.

Synapse – a structure that links (through a synaptic cleft) a nerve cell with another nerve or some other cell type. In an electrical synapse, an activated neuron develops an electric signal on its presynaptic cell membrane, which induces a voltage change on the postsynaptic cell. In a chemical synapse, electrical activity on the presynaptic cell membrane initiates the release of a neurotransmitter and its diffusion to a postsynaptic cell.

Texture : a characteristic of a substance or a material. Texture exists all around us. It can be actual (natural, invented, or manufactured) or simulated (made to look rough, smooth, hard or soft, or like a natural texture). Simulated textures are made to represent real textures, such as smooth arms or rough rock formations. But they are not actual textures; if you touch the picture, you feel only the paint or the pen or pencil marks.

URL means Uniform Resource Locator. Also, resources are called well-known codes, such as www.sciam.com/, used in hyperlinks (activated by clicking).

Variable is a feature or a factor that can take on different values.

Vernacular is an adjective to describe the local, indigenous, folkloristic, common, or simple form of a specific culture, architecture, language, or other human activities or product.

Vernacular language is typical of a particular country or region or a **dialect** that can be regional, professional, or otherwise linking a group of people. It may also denote the architecture of houses in a particular region.

Virtual reality (VR)   refers to immersive experiences created with the use of real-world content (e.g., 360° video), purely synthetic content (for instance, computer generated), or a hybrid of both.

Visuals   are pictures, diagrams, graphics, film clips, or various forms of display.

Visual media   are primarily images, videos, and infographics. PowerPoint presentations may serve as an example.

Visualization   means the communication of information with graphical representations. Interactive visual representations of abstract data use easy-to-recognize objects connected through well-defined relations.

Wireframe   is a skeletal three-dimensional model in which only lines and vertices are represented. It can also be a way to design a website at the early structural level, a 2D outline of a web page or app that displays the functional elements of a website.

Index

Note: Page numbers in *italics* indicate a figure and page numbers in **bold** indicate a table on the corresponding page.